# SOLARSPACES

How (and Why) to Add a Greenhouse, Sunspace, or Solarium to Your Home

## DARRYL J. STRICKLER

VAN NOSTRAND REINHOLD COMPANY
New York Cincinnati Toronto London Melbourne

*This work is dedicated to a bright future for all of us*
*and especially to*
*Lauren, Kyle, Derek, Kristin, and Trevor,*
*who have lighted my path.*

Printed in the United States of America

Designed by Ginger Legato

Published by Van Nostrand Reinhold Company Inc.
135 West 50th Street
New York, New York 10020

Van Nostrand Reinhold
480 Latrobe Street
Melbourne, Victoria 3000, Australia

Van Nostrand Reinhold Company Limited
Molly Millars Lane
Wokingham, Berkshire RG11 2PY, England

16 15 14 13 12 11 10 9 8 7 6 5 4 3 2 1

**Library of Congress Cataloging in Publication Data**

Strickler, Darryl J.
    Solarspaces.

    Bibliography: p. 136
    Includes index.
    1. Solar heating.  2. Solar greenhouses.  3. Garden rooms.  4. Porches.  I. Title.
TH7413.S773 1983        690′.89        82–15959
ISBN 0-442-27803-9
ISBN 0-442-27802-0 (pbk.)

# Contents

*Frontispiece, 1–1. Enclosed atrium (Designer: Stephen Merdler, Santa Fe, NM; photo: Darryl J. Strickler)*

# ONE

# Solarspaces: An Introduction

*What is a solarspace, and why would you want to own one?*

A solarspace is a part of a home or other building that can provide supplementary heat and additional floor space, as well as a place to grow food or houseplants. If properly designed and oriented toward the south, a solarspace can be used year-round and can even assist with the cooling of the house in summer. Such a space can be added to an existing house, or it can be integrated into the design of a new house. Solarspaces can be designed to serve a specific function or a variety of functions, as well as to suit the overall characteristics of a specific house and the tastes and financial means of individual homeowners. Depending on its design and who builds it, a solarspace can range in cost from as little as five dollars or less a square foot to as much as fifty dollars or more a square foot.

*Solarspace*, as used in this book, is a generic term that includes attached solar greenhouses, sunspaces, solaria, and enclosed porches. A solarspace built primarily for food production, including fruits, vegetables, herbs, and fish, might be referred to as an attached solar greenhouse. The words *attached* and *solar* distinguish this type of greenhouse from freestanding, all-glass, energy-inefficient commercial greenhouses. An attached solar greenhouse (or just ''greenhouse,'' as it is referred to in this book), would be designed somewhat differently from what might be called a sunspace. The primary purpose of a sunspace is to add floor area to a house; production of heat and food are of secondary importance. A sunspace more than one story tall is sometimes called a solarium, or if it is in the center of a house, an enclosed atrium. Regardless of what you choose to call a particular type of solarspace, such a space should be designed to fit the needs of the people who will use it and the characteristics of the house of which it is a part.

Thousands of homes throughout the world have attached or integrated solarspaces. Although the concept of solarspaces has been around for centuries in the form of sun-rooms, sun porches, and attached ''hobby'' greenhouses, the majority of existing solarspaces

1-2. Attached solar greenhouse (Designed and built by The Solar Project, Lancaster, PA; photo: David S. Strickler, Strix Pix)

1-3. Sunspace (Hedrick residence; photo: Darryl J. Strickler)

in America have been built since the mid-1970s. The recent rediscovery, and in fact, reinvention of the solarspace can no doubt be attributed to the oil embargo of 1973, the increasing cost of utility-supplied energy, and the realization that even food production and distribution depends upon nonrenewable forms of energy. Moreover, with recent deregulation of petroleum and natural gas prices, we can expect the costs of all forms of energy, including electricity and even wood, to increase at a rate of 20 to 30 percent a year. (The National Association of Home Builders Research Foundation estimates that the cost of natural gas will increase 430 percent between 1982 and 1990.) We can also no longer be certain that fruits and vegetables grown in distant parts of the country will arrive fresh at our local food store when thousands of gallons of petroleum are needed to transport them.

Instead of merely speculating on the grim possibilities of energy and food shortages, many people have decided to act. Such action has taken a variety of forms, from a headlong search for more oil and natural gas, to the development of alternative energy and food sources. Much of this activity is focused on ways to improve centralized supply and distribution networks; other efforts have been directed toward decentralization. The addition of a solarspace that can provide both heat and food to the occupants of a house is an outstanding example of one approach to decentralization of the supply and distribution of energy and food.

Solarspaces have become the most popular solar retrofit option (aside for solar water heating) because a solarspace can be added to almost any existing structure that has a south-facing wall and because a solarspace can be designed to fit virtually any house, budget, or life-style. Solarspaces are now appearing on many newly constructed houses, including manufactured and precut homes. Literally hundreds of manufacturers also offer prefabricated sunspaces or greenhouses that can be attached to a new or existing house.

Many people who have a greenhouse, sunspace, solarium, or enclosed porch on their home have set out with the modest goal of reducing their dependence on the power grid and the food supply network; others have aimed for nothing short of full energy independence—cutting the wires or gas lines, as it were, and taking a detour around the produce section of their local food store. At the other end of the spectrum are people who build solarspaces in what appears to be an effort to participate in a sort of "solar chic" that runs through some circles of society and architectural design.

The reasons for building a solarspace vary as widely as the forms such a space can take and are as diverse as the population itself. For example, low-income people and others on fixed incomes have built them literally to survive. At the same time, more affluent homeowners in all parts of the country have built solarspaces primarily because they are attractive. Whatever their reasons, people who have a solarspace in their home have, in most cases, made a wise investment in their own future. The return on the investment will depend on how effectively they use the space, how rapidly the cost of conventional energy increases, and how much they spent on their solarspace. As you will discover by reading this book, the cost of a solarspace depends on who builds it, and on the quantity and quality of materials used.

An inexpensive greenhouse can pay for itself in a few years by supplying both heat and food to its owners. It can also be used to grow plants or food that can be sold to others, thereby shortening the payback period. By the same token, a well-designed sunspace or solarium can not only provide heat, food, and additional space, but can increase the value of the property if that solarspace is attractively finished and in harmony with the rest of the house. Most important, a solarspace can enhance the lives of people who use it.

If you are not convinced that you want to build a solarspace in your home, you may find the following questions helpful. How would you like to have a warm, sunny place to spend a cold winter afternoon? Wouldn't it be a real treat to "dine out" amidst a lush, plant-filled environment without leaving home? What about digging your hands into the fertile soil of your greenhouse after a long day in "the jungle"? How does a red, ripe tomato from your own greenhouse in the dead of winter sound, as compared to the greenish pink plastic variety sold in food stores at that time of the year? What about staying warm during a power blackout on a cold night? Would you appreciate a homegrown dinner of fresh fish, flavored with herbs and lemon, and a salad made from fresh greens? Wouldn't you be more comfortable in the winter if the air in your house were not so dry because the plants in your sunspace add moisture to the air? Would you feel more like making (or eating) breakfast if you could watch the sun rise behind the plants in your greenhouse outside the kitchen window? What about ending your day with a good soak in a tub of solar-heated water in your solarium? How would you feel about adding space to your house if that space also reduced your fuel bills? How would making better use of the limited resources of this planet affect the way you view yourself, your relationship with nature, and your relationship to the utility company? Many questions, with many possible answers. Perhaps

1-4. Food and heat, benefits of an attached solar greenhouse (Lerner Greenhouse; photo: Darryl J. Strickler)

1-5. A well-designed solarspace can increase the value of the house to which it is attached (Crawford Residence; solarspace designed and built by Eric Sorenson; photo: Darryl J. Strickler)

you have found among them a good reason to start planning a solarspace for your home.

## REAL PEOPLE; REAL SOLARSPACES

You may find it helpful at this point to look at some examples of existing solarspaces in various parts of the United States. The following stories about people and their solarspaces are intended to provide food for thought as well as some ideas of how you might plan and build your own solarspace.

## Riccardo and Regina Pacini
## Denver, Colorado

In 1977 Riccardo Pacini purchased an eighty-year-old Victorian-style house. Although the masonry house was solidly built, it needed some refurbishing. Because he planned to rent the upstairs apartment, Riccardo began to think about how he could add space to the downstairs. He also began to think about how he could create some of the flavor of his native Italy in the heart of Denver, Colorado. Riccardo discussed his plan with Regina Taylor, who later became his wife (they were married in their solarium). At the time she was studying drafting and design and needed an outlet for her considerable creative talents as a designer. Their discussions resulted in a plan to build a thirty-eight- by eleven-foot solarium addition to the south side of the Pacini residence. The solarium would not only add living space, but it would also offer Riccardo space to grow plants. Being a person who likes to take charge of his own environment, Riccardo set about the task of constructing the solarium with a little help from a friend in September, 1980. Three months and $8,500 later, the Pacini solarium was completed.

The fruits (and vegetables) of labor has become a literal term in the Pacini household these days—tomatoes, beans (growing up the spiral staircase), squash, dwarf fruit trees, tree roses, fig trees, and houseplants. Lest you get

1-7. Pacini solarium viewed from backyard (Photo: Darryl J. Strickler)

the impression that Riccardo's only interest was in creating a lush subtropical environment reminiscent of Italy in the heart of the American West—which is something he certainly succeeded in doing—you need only ask to inspect his charts that show the energy consumption of his house. His use of natural gas for heating purposes was reduced by two-thirds after the solarium was completed.

To gain a more lasting impression of the value of the solarium, you might ask Burrito, the member of the Paccini household who lives in it on a full-time basis. But then, you probably cannot count on a parrot to give you an unbiased report.

1-6. Pacinis' solarium provides a pleasant retreat for the family (Photo: Darryl J. Strickler)

# Vern and Jackie Weiss
# Ashland, Oregon

Vern and Jackie Weiss had been thinking about building an attached solar greenhouse for a number of years. In the spring of 1980, a neighbor who knew about their interest told them she had heard a radio announcement for a greenhouse construction workshop sponsored by SUNERGI, a solar energy group in Ashland. Vern's phone call to SUNERGI resulted in a site visit to the Weiss residence by Fred Gant and Scott Cummings. After attending a workshop at another greenhouse construction site, the Weisses completed their plan with designer Scott Cummings. The south wall of the ten- by eighteen-foot Weiss greenhouse was to be glazed with six double insulated patio door glass replacement units. Six of these units were also used as roof glazing to collect the diffuse sky

1-8. The Weiss greenhouse shaded by deciduous vines (Photo: Darryl J. Strickler)

radiation associated with cloudy Oregon winters. With the plans drawn up, Vern and Jackie ordered the materials and prepared the foundation for their greenhouse. On a cool Saturday morning of Memorial Day weekend in 1980, six volunteers showed up bright and early to begin building the Weiss greenhouse. As the long weekend wore on, the greenhouse began to take shape. By Monday afternoon the greenhouse was fully enclosed and essentially complete; only the interior finishing work was left to be

done by Vern and Jackie. After a few more weekends of diligent effort, the entire project was finished. The total cost of the project, including feeding the volunteer workers, came to $2,800.00. By claiming tax credits of $700.00,

1-9. Weiss greenhouse with shade cloth over roof glazing (Photo: Darryl J. Strickler)

the Weisses reduced the net cost of their greenhouse to $2,100.00 ($11.66 per square foot), which they view as a real bargain considering the 40 percent reduction in the cost of heating their home. Although their original motivation for building a greenhouse was to reduce the cost of heating their totally electric home, Vern and Jackie Weiss have derived many other benefits from their greenhouse. In the heating and growing seasons since its completion, the Weiss greenhouse has provided a comfortable home for a wide variety of houseplants and vegetables.

# Charles and Eileen Dunnell
# Frederick, Maryland

Charles and Eileen Dunnell have had a long-standing interest in energy conservation and organic gardening. These interests, along with their appreciation of early Federal-style architecture, influenced their purchase of a two-story brick house in Frederick, Maryland, built in 1907. Although the property is in a residential section of the city, adjacent to the downtown area, it provided adequate space for gardening and, of equal importance to the Dunnells, space to build a solar addition on the south side of the house.

After purchasing the property in 1976, Charles began reading everything he could find on the design and construction of solar additions—which, at the time, was not very much. His planning efforts resulted in a fourteen-foot-tall sunspace built in 1981 by Kent Briddell, a solar designer and builder from the Frederick area. The total plan to retrofit the house included adding insulation to the exterior of the north, east, and west walls of the house, an active solar hot water system, and the twenty-six-by ten-foot sunspace. The total cost of the sunspace addition was about $6,200.00 ($23.85 per square foot), excluding the solar hot water system.

1-11. Dining area in Dunnells' solarspace (Photo: Darryl J. Strickler)

The Dunnell sunspace includes a small deck off the kitchen for "dining out" and ample space and planting beds for growing houseplants and vegetables. Charles and Eileen use the sunspace to extend the growing season in their lush outdoor vegetable garden. To help shade the sunspace in summer, Charles uses a variety of fast-growing runner beans that cover the glazing surface.

1-10. Dunnell solarspace with collector panels (Photo: Darryl J. Strickler)

## Ethel Smith
## New Holland, Pennsylvania

When a friend told Ethel Smith about The Solar Project, operated by the Lancaster County (PA) Community Action Program, she was eager to investigate their services. Although it seemed hard to believe, Mike Carey of the Solar Project told her that if she qualified and if her house was suitable for such a structure, the project staff would design and build an attached greenhouse/ sunspace on her house without any cost to her. Finding it difficult to refuse such a deal, Mrs. Smith agreed to have the structure built. The ten- by twenty-four-foot structure, designed by Mike Verwey, was completed in May, 1981, at a cost of $2,000.00 ($8.33 per square foot) for materials only.

To discover whether Mrs. Smith appreciates her greenhouse/sunspace, you need only look at the profusion of healthy plants and the neatly arranged dining space. She uses her solarspace to give the vegetables and flowers she will later transplant outdoors a head start in early spring. Although she heats her house with wood, Mrs. Smith appreciates that her solar-

*1-12. Ethel Smith's solarspace (Designed and built by The Solar Project, Lancaster, PA; photo: David S. Strickler, Strix Pix)*

space provides additional heat—as well as a warm place to spend a sunny winter afternoon—and helps to keep her house cooler in summer. The greenhouse is shaded in summer by a large tree to the west and has a well-designed ventilation system to keep it from overheating.

*1-13. Smith solarspace in summer cooling mode: roof vents and stem wall vents open; screen door provides additional airflow (Photo: David S. Strickler, Strix Pix)*

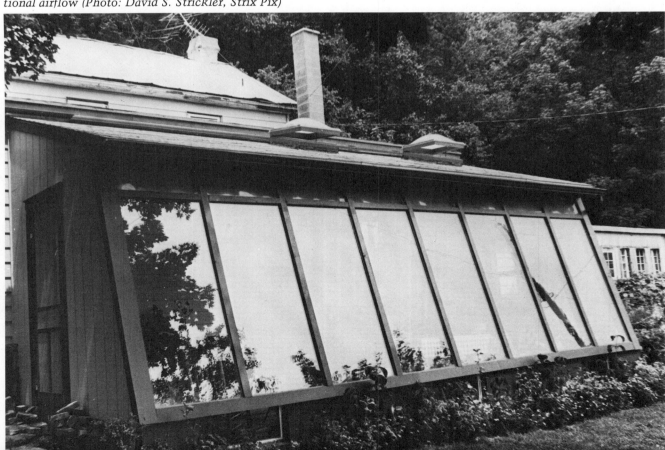

## Jim and Joyce Baker
## Portland, Oregon

Instead of moving to a larger house with a higher mortgage interest rate, Jim and Joyce Baker decided to add a room to their house that would provide a place to relax and entertain friends. Rather than a more predictable family-room addition, the Bakers wanted a solarspace that would include a sunken hot tub as its focal point. A visit to friends in Canada who have a hot tub had convinced them that a hot tub would be a worthwhile investment, not only providing relaxation, but also improving the value of their home. Because they wanted to use the tub year-round and take advantage of free solar energy, they decided to build a solarspace to enclose the tub.

Their busy schedule did not allow the Bakers to build the solarspace themselves, so they employed the Neil Kelly remodeling firm in Portland to build it for them. David Leach, then a designer for the firm, came up with a plan for a twelve- by fourteen-foot solar room that would provide the kind of indoor environment the Bakers were seeking, along with supplementary heat for the house and a private sun deck on the roof of the solarspace. When the addition was completed in the summer of 1981, the Bakers and their daughter, Jennifer, did not waste any time putting it to use. In fact, when Jennifer's friends come over to visit, they automatically bring their swimming suits so they can enjoy a good soak in the Bakers' hot tub.

In the basement of the Bakers' house is a storage system with a heat exchanger that captures and distributes solar heat on sunny days—which, incidentally, are more plentiful, especially in the spring and fall, than is generally believed by people who do not live in Portland. Although their investment of $28,000 was considerable, the price included new wiring, the hot tub and related equipment, and an active-type solar hot water system. Oregon's solar incentive program, and federal energy tax credits have helped to reduce the net cost of the Bakers' solarspace.

1-14. Sunken hot tub provides the focal point for the Bakers' sunspace (Photo: Darryl J. Strickler)

1-15. Bakers' sunspace with deck on roof (Photo: Darryl J. Strickler)

# Denzil Surrett
## Sandy Mush, North Carolina

During the winter of 1979 to 1980, octogenarian Denzil Surrett and his son, Clifford—himself a senior citizen—spent more than six hundred dollars for the fuel oil used to heat only two rooms of their home. The drafty old farmhouse they lived in at the time was built in 1884—after the Civil War and before insulation. Although Denzil and Clifford still live in the same old house, it is no longer drafty and no longer costs as much to heat. Thanks to the efforts Paul Gallimore and the staff of the Long Branch Environmental Education Center in Leicester, North Carolina, the Surretts' house is now weatherized and has a solarspace on the front of it. As a result of the weatherization and the heat provided by the solarspace, the Surretts now spend about two hundred dollars a year on fuel oil to heat their home in the cold climate of North Carolina's Newfound Mountains.

The Surrett retrofit project began after Denzil visited another solarspace built by the Long

1-17. The completed Surrett solarspace (Photo: Paul Gallimore, Long Branch Environmental Education Center, Leicester, NC)

Branch Environmental Education Center. After carefully inspecting the attached solar greenhouse on a staff residence at the center, Denzil announced: ''I just got one thing again' it.'' ''What's that?'' he was asked. Denzil's response: ''I ain't got one.''

The Surrett solarspace has served as a model for the Sandy Mush community. Because it is highly visible along a well-maintained state road, hundreds of people have visited the Surretts to talk with them about their solarspace. Denzil and Clifford use the solarspace to grow a variety of houseplants and flowers, as well as tomatoes, green peppers, broccoli, and cabbage throughout the cold winters of their mountain region.

The story-and-a-half solarspace, which was glazed with recycled, tempered glass, encloses the old front porch of the residence. The concrete steps of the porch and water containers serve as thermal mass to store heat. The total square footage, including the porch floor and the floor of the ten- by sixteen-foot solarspace is 280 square feet. Materials for the Surrett project, which was completed in 1980, cost about $1,400 ($5 per square foot). The overall success and low cost of the project is largely the result of the ingenuity and hard work of Paul Gallimore, the Long Branch Environmental Education Center staff, the Surretts, and volunteers who helped build the solarspace.

1-16. Denzil and Clifford Surrett and volunteers about to begin building the solarspace (Photo: Paul Gallimore, Long Branch Environmental Education Center, Leicester, NC)

## Elizabeth and Carl Gilmore
## Taos, New Mexico

Elizabeth and Carl Gilmore are not what you would call ordinary people, so it should come as no surprise that they designed an extraordinary solarspace for their home. Elizabeth is a nationally known midwife who has ushered more than four hundred people into the world. In addition, she has somehow found time to write numerous articles for journals and magazines on the subject of home births. Carl is a self-employed appraiser and investment counselor. Because both Gilmores used their home as an office, they needed additional space for working and meeting with their respective clients, as well as more room for family activities.

Carl and Elizabeth began planning their solarspace in 1979. Since the south wall of their house faced more than 20 degrees east of south, they decided to include angled, south-facing skylights in the roof of their solarspace. To insulate the large skylights at night and during cloudy periods in winter, and to block the summer sun, the Gilmores selected Zomeworks Skylids—aluminum louvers, or foils, that open and close automatically (without electricity) as a result of changes in density of Freon gas in the canisters attached to the louvers.

With their planning efforts nearly complete, the Gilmores lined up contractor Robin Taylor and master carpenter Peter Barlow to build their solarspace in 1979. Although the thirteen- by thirty-foot space cost, in Elizabeth's words, ''a lot'' (just over $20,000, or about $52 a square foot), it has not been difficult for the Gilmores to justify its expense in relation to the enjoyment and solar heat they have derived from it.

The interior of the Gilmore solarspace is an inviting and visually attractive place, furnished with a comfortable sofa, other furniture, and a wood-burning stove, as well as a redwood-decked hot tub that is partitioned from the rest of the space. The tub room includes an etched-glass panel in the south wall made by one of the Gilmores' neighbors, Charles Collins.

The fine workmanship and detailing is everywhere in evidence inside and outside the solarspace. Apparently other visitors have also come away with the feeling that the total visual,

1-18. The fine detailing by contractor Robin Taylor and carpenter Peter Barlow is evident throughout the Gilmores' solarspace (Photo: Shirleen Strickler)

1-19. Elizabeth Gilmore, midwife, author, and proud solarspace owner (Photo: Darryl J. Strickler)

sensory, and extrasensory effect of the space is greater than can be attributed to the sum of its parts. (I suspect it has as much to do with the Gilmores themselves as it does with anything else.) Houseguests ask to sleep in the solarspace; the Gilmore children, Anna, Ezra, and Jan, inhabit it regularly, summer and winter; and the family gathers there for breakfast. In short, everyone senses that it is a magical place.

# Jeff Milstein
# Woodstock, New York

Jeff Milstein is an architect and writer who has specialized in passive solar design since 1975. He coauthored *Designing Houses: An Illustrated Guide To Building Your Own House* (Overlook Press, 1976) and has written numerous articles and reviews for *Popular Science*, including an article in the May, 1982, issue that describes the solarspace featured here. One of Jeff's professional goals has been to make well-designed projects more accessible to do-it-yourselfers. His own solarspace, which he built with the help of friends and some hired labor, is a testimony to Jeff's do-it-yourself concept.

All of the materials for Jeff's nine- by twenty-foot solarspace, built in the fall of 1981, cost about $4,256.00 ($23.64 a square foot). Although his choice of an insulated sliding glass door and semicircular and triangular glazing in the east wall of Jeff's solarspace add considerable visual impact, Jeff indicated that the cost of his solarspace could have been reduced if these features were excluded. The use of a patterned or stained concrete floor in place of the quarry tile he used would result in further savings of about $500.00. Jeff figures that an economy version of his solarspace could be built for between $2,700.00 and $3,000.00.

Although the post-type foundation Jeff used for his solarspace is less expensive and easier to construct than a typical concrete foundation—especially in regions where the frost line is three to four feet deep—Jeff stated that his design could be easily adapted to a typical foundation. (The raised floor of Jeff's solarspace is insulated to R–38 beneath the quarry tile surface.)

The special features of Jeff's solarspace include a portable Jacuzzi that is used outdoors in summer and moved into the solarspace in winter. Jeff and his friend Christine spend many winter evenings relaxing in the Jacuzzi and otherwise enjoying each other's company in the solarspace. In addition, Jeff does much of his writing and design work in his solarspace.

A thermostatically controlled, reversible vent-axia fan draws warm air from the space during sunny winter days. The same fan reverses to return warm air from the house to the solarspace during extremely cold nights to keep plants in the solarspace from freezing. The well-designed ventilation scheme that includes wood-frame awning windows as low vents, an operable skylight, and two fourteen-inch wind turbines as exhaust vents keeps the space comfortable during upstate New York summers.

(Do-it-yourself plans for Jeff's solarspace, including construction details, materials list, and building tips are available. See Appendix for ordering information.)

*1-20. Milstein solarspace, Woodstock, NY (Photo: Jeff Milstein)*

*1-21. Interior of the Milstein solarspace (Photo: Jeff Milstein)*

## Kathy and Doug Kelly
## Woodinville, Washington

In 1977 Doug and Kathy Kelly purchased an attractive two-story home in the Seattle metropolitan area. They liked everything about the home except its blank, windowless south wall that tended to keep the house dark during the daytime. They had been thinking about doing something with the south end of their house for a number of years. In February, 1981, they attended a do-it-yourself fair and saw a full-scale model of a solarspace designed and built by Tim Magee, Rollin Francisco, and the staff of Rainshadow Design in Seattle.

1–23. Sun-drenched interior of the Kellys' solarspace (Photo: Darryl J. Strickler)

1–22. The Kellys' solarspace offers an excellent alternative to a blank wall (Photo: Darryl J. Strickler)

It did not take the Kellys long to decide that an attractive solarspace by Rainshadow was exactly what they needed for their barren south wall. Not only would it provide heat, a place for Kathy's plants, and extra space, but the French doors that would connect it to the house would admit much needed light to living areas. The Kellys did not waste any time contracting Rainshadow to design and build their solarspace. The attractively finished ten- by eighteen-foot structure was completed in April, 1981, at a cost of $8,600.00 ($47.77 per square foot). The clean lines and tile floor of the solarspace make it a favorite place for relaxing and ''dining out.'' The

Kellys' sons, D.J. and Jason, have even been known to do their homework in the space on occasion.

A thermostatically controlled fan between the solarspace and the house is used to extract solar-heated air. The Kellys report that the fan (set at 72°F) sometimes operates as late as 9:00 P.M. on spring and fall days when heat is required for the house. In summer a fan at the top of the west wall of the solarspace exhausts hot air to the outdoors. French doors in the east wall of the solarspace and a window in the west wall provide good cross ventilation for summer cooling.

The Kellys are extremely enthusiastic about their solarspace and feel that it has increased the value of their property beyond the actual cost of the solarspace. This is especially true since the net cost of their solarspace was reduced to about $3,000 as a result of state and federal income tax credits.

## John and Ann Rodgers
## Santa Fe, New Mexico

1-25. Rodgerses' solarspace viewed from southwest (Photo: Darryl J. Strickler)

Shortly after Ann and John Rodgers purchased a forty-year-old adobe house in 1978, they began thinking about adding space so their then toddler-age son, Paul, would have a place to do his "work" —usually known as play for people less than five. Ann, who was then a nursery school teacher, was also considering the possibility of opening a day-care center in the Rodgerses' home. With the help of a solar designer-builder, Charles Loesch, the Rodgerses came up with a plan for an eighteen- by thirty-five-foot solarspace that incorporated an air lock entry to the house, a bathroom, utility and storage areas, and a dining area. Construction began in the spring of 1979 and was completed about three months and $20,000.00 later. The cost of $31.75 a square foot was a remarkable value considering how the Rodgerses have since used their solarspace. Moreover, that cost also included extra plumbing, a gray-water system, and bathroom fixtures.

Not many people use their solarspace for as many purposes as the Rodgerses use theirs. From the beginning the space has served as Paul's play area and the home of the family's solar cats. One-third of the planter box in the Rodgerses' solarspace is a sandbox—intended for

1-24. Air lock entry (through solarspace) to the Rodgerses' residence (Photo: Shirleen Strickler)

Paul. In addition to dining in their solarspace, the Rodgerses do their family wash in the solarspace utility area and hang it up to dry on a clothesline stretched across the space. The solarspace makes an excellent clothes dryer, and the herbs, spices, lettuce, Swiss chard, beans, tomatoes, and flowers grown in the solarspace certainly appreciate the extra moisture given off by the wet clothes. Ann, who is now a stained glass artist, finds that the solarspace is an excellent place to display her work.

The sloped south glazing and clerestory windows in the roof admit adequate sunlight to illuminate and heat both the solarspace and the house on sunny winter days. A wood-burning stove in the solarspace is used on extremely cold nights to keep the Rodgerses, their plants, and their solar cats comfortable. Along with many other solarspace owners, the Rodgerses have not yet found a satisfactory scheme for insulating the sloped glazing of their solarspace at night, but they plan to add a movable insulation system. They have tried several strategies for summer shading, including growing sunflowers in the planting beds outside the south wall of their solarspace. The *vigas* logs that extend out horizontally above the glazing were included to provide a framework to hold shade cloth.

A final note of interest about this solarspace comes as a result of John's work as a health physicist at Los Alamos Scientific Lab. As a part of his research on indoor air pollution, John has monitored the presence of radon, a radioactive gas emitted from the building materials and earth inside his own tightly built solarspace. He discovered that the air in his solarspace contained forty times as much radon as the outdoor air. Although he does not consider this to be a dangerously high level, John recommends the use of an air-to-air heat exchanger in tightly sealed buildings (see page 35 and chapter 3). He is considering installing one in his own solarspace.

## Bob and Betty Flynn
## Auburn, Alabama

What would you do if you needed to seat twenty-two people for a family Christmas dinner and your dining room was not large enough to hold everyone? For Bob and Betty Flynn, the answer came easily: use their solarspace as a dining room. Thus it was that the Flynn clan gathered for Christmas dinner amidst the many poinsettias that are year-round residents of the Flynns' solarspace.

The Flynns' decision to build their solarspace in 1981 was a natural extension of their previous efforts to reduce the energy consumption of their totally electric home. Their conservation efforts, which included caulking and weather-stripping and adding insulation and storm windows, began in 1977 when they purchased their present home. Bob, who is a retired army officer, has kept careful records of the number of kilowatt hours (KWH) of electricity the house has used since 1978. Although his figures do not account for varying weather conditions, the Flynns used 2,492 KWH in March, 1981, (before the solarspace was completed) and only 1,482 KWH in March, 1982, a reduction of approximately 40 percent.

Both Betty and Bob refer to their "solar room" with a great deal of pride and enthusiasm. They not only appreciate the energy savings, but also thoroughly enjoy the time they spend caring for, and "being with," their many houseplants. Because their solarspace is built into what was the open part of the back of their U-shaped house, the heat produced in the solarspace can make a loop through the entire house via open windows and doors to the house on sunny winter days. The solarspace also reduces the heating and cooling load of the house by protecting three previously exposed exterior walls. Beneath the floor of the Flynn solarspace is a fan-charged rock bed that helps keep the solarspace warm on winter nights, thereby further reducing heat loss through the walls of the house. The top surface of the solarspace floor is loose gravel, which provides some heat storage and allows water to drain away easily.

For the Flynns and many other homeowners in warmer regions of the country, the problem of

1-27. The Flynns' solarspace, with frame-mounted shade cloth on exterior of glazing (Photo: courtesy of Phifer Wire Products, Tuscaloosa, AL)

cooling their house in summer is as much of a concern as heating it in winter. A solarspace that substantially increases the cooling load of a house is of limited value. Fortunately, the Flynns do not have this problem. Thanks to the well-designed ventilation system and frame-mounted shade cloth on the exterior of the glazing surface, the Flynns' solarspace has already demonstrated that it can handle just about anything an Alabama summer can dish out and still remain comfortable for people and plants.

The Flynns' twelve- by thirty-foot solarspace was designed and built by Jerry Liveoak at a cost of $9,600.00—$26.66 per square foot. At last report Bob and Betty were planning to apply for state and federal tax credits totaling approximately $6,500.00, which would reduce the net cost of their solarspace to $3,100.00—quite a bargain when you consider the energy savings and enjoyment the Flynns have already derived from it.

## Eric Asendorf and Martha Mortensen West Falmouth, Massachusetts

Eric Asendorf moved to Cape Cod from Baltimore, Maryland, in 1970. At about the same time, Martha Mortensen moved to the Cape from Carlisle, Massachusetts, to assume a teaching position. Eric and Martha met at the elementary school where they both taught.

In 1978 Eric purchased the traditional cedar-shingled summer cottage that is now Eric and Martha's year-round residence. The task of winterizing the cottage—which came complete with no heating system and no insulation—was begun in earnest with the addition of storm windows, weather stripping, insulation, and a wood-burning stove. With the house tightly sealed and insulated, it was time to consider adding solar features.

Because of the character and charm of the exterior of the home had been faithfully preserved, it was very important that any solar features be in keeping with its classic Dutch Colonial lines. With the help of Bob Skilton and the staff of Weather Energy Systems, a solar design and construction firm in Pocasset, Massachusetts, an eight- by twenty-four-foot solarspace with sloped glazing was built. The solarspace would not only provide supplementary heat and additional space, but would also blend

1-29. Well-integrated solarspace on Mortensen/Asendorf residence (Photo: James Marlinski, © 1982; courtesy of Weather Energy Systems, Pocasset, MA)

well with the existing lines of the house. At the time (1979), Weather Energy was developing its now perfected Sun Haus glazing gasket system and its Plexus fan control system for solarspaces. Eric and Martha's solarspace was among the first built by Weather Energy that incorporated these glazing and fan systems.

Construction of the solarspace was completed during the winter of 1979 at a cost of $10,000 ($52 per square foot). The net cost of the solarspace was reduced, however, to around $5,000 ($26 a square foot) as a result of federal and state tax credits. In addition, the solarspace will not add to Eric and Martha's real estate taxes until 1999 by virtue of a Massachusetts incentive program that postpones assessment of the added value of solar features for a period of twenty years.

Martha and Eric are pleased with the aesthetic appeal and function of their solarspace. The added room provides a pleasant retreat on sunny days. With its automated fan system, the solarspace also aids significantly in cooling and ventilating the house in summer. The Plexus fan between the house and solarspace extracts warm air and moisture from the house in summer, which helps to reduce the interior temperature of the house and prevents mildew growth—a common problem in houses on Cape Cod.

As times and people change, careers change; so do methods of home heating. Martha is now a yacht broker; Eric is a stockbroker. Both are in their early thirties. Not too many years ago when they were teaching elementary school children and living in their first apartment, the amount of nonrenewable energy they used was determined by the dial of a thermostat. Now Eric and Martha depend on wood and the sun to get them through the long winters on Cape Cod.

1-28. Interior of Asendorf/Mortensen solarspace (Photo: James Marlinski, © 1982; courtesy of Weather Energy Systems, Pocasset, MA)

## Elizabeth and Glenn Saccone
## Broken Arrow, Oklahoma

Elizabeth and Glenn Saccone's solarspace neatly demonstrates a couple of things. First, a solarspace need not look like a solarspace to be a solarspace. Second, their solarspace proves that people who never built anything before can at least handle the finish work, if not the entire task of constructing a solarspace.

When the Saccones moved to Oklahoma from the Midwest, they brought with them an energy-conservation ethic developed during the long winters spent in Chicago and Kansas City. They soon discovered that not all of their new neighbors and friends shared their views on the need to conserve—perhaps because of the abundant energy resources in Oklahoma. They were by no means discouraged by other people's views on the feasibility of using solar energy for home heating. Because they knew that sunshine is also an abundant energy resource in the Sooner state, they felt an even greater need to pursue their plans.

Although they originally considered adding an all-glass, aluminum-frame, prefabricated greenhouse to their house, they soon discovered through their reading and a consultation with Jim Johnson of Johnson and Sun that an all-glass structure would neither fit their expectations nor improve the value of their house. Together, Johnson and the Saccones designed the well-integrated twelve- by eighteen-foot structure that now graces the south side of the Saccone residence.

Because the height of the south wall of Saccones' house did not permit a more typical shed-type solarspace, Johnson planned to join a gable-roofed solarspace with the existing south roof of the house. The Saccones wanted to keep the cost of their project as low as possible, but they had neither the time nor the experience needed to take on the entire project themselves. They felt, however, that they could at least install the insulation, do the finish work, and make their own thermal shades for the glazing areas of the solarspace. They hired Johnson to rough in the structure and install the glazing, wiring, roofing, and siding at a cost of $7,200 for labor and materials. The cost of the remaining materials the Saccones needed to complete their solarspace came to about $2,100, making the total

1–30. A well-integrated solarspace: the Saccone residence (Photo: Darryl J. Strickler)

1-31. Interior of Saccone solarspace (Photo: Darryl J. Strickler)

cost of the project $9,300 (about $43 a square foot). The Saccones' finish work, including a dry-laid brick floor over leveled sand, added about $10 a square foot to the rough-in price of $33 a square foot. Their net cost, after state and federal tax credits, was roughly $3,000 (about $14 per square foot).

To construct the shades that are rolled down over the glazing of their solarspace on cold nights and during the summer, the Saccones laminated fabric on Foylon 7194 with fabric glue and used a porch awning–type pulley-and-rope system and magnetic tape for edge seals. Materials for their radiant shades, including all the hardware, cost about $1.83 per square foot of glazing covered by the shades.

At the time of this writing, the Saccones' solarspace was brand new and as yet untested. Considering the design of its heating and cooling features, however, there is every reason to believe it will meet the realistically high expectations they have for it. Whatever else it does, the Saccones believe their new solarspace has already increased the value of their house beyond what they actually spent for the solarspace.

## Teresa and Victor Vigil
## San Luis, Colorado

Anyone can have a solarspace, right? All you need is lots of money. Wrong. Victor and Teresa Vigil spent only $500.00 on their eleven- by eighteen-foot attached solar greenhouse ($2.53 per square foot). With the help of Arnold and Maria Valdez of the People's Alternative Energy Service in San Luis, a $200.00 grant from the ARCA Foundation, and lots of determination and hard work, the Vigils constructed their greenhouse—even though they had never built anything before. This was a remarkable feat indeed, but what is even more remarkable is the productivity of the greenhouse—or, more accurately, of Teresa Vigil.

Not too many people in the world can work the kind of magic Teresa does when it comes to growing things. Even fewer can make it work year-round, season after season, in an attached solar greenhouse. Her secret? Experimentation—finding out what works best—and lots of tender, loving care. To make their greenhouse soil, the Vigils gathered rich compost, leaves, and decomposed pine needles from around the base of trees and mixed it with manure and what is known as *tierra negro* (black dirt) in the San Luis Valley.

*1-33. Vigil greenhouse lit by morning sun (Photo: Shirleen Strickler)*

The first tomato plant Teresa started in her greenhouse grew to a height of twelve feet—not a bad start. In addition to tomatoes, Teresa has basil, parsley, leaf lettuce, radishes, and a peach tree growing in her greenhouse. She also grows marigolds, which she takes to sick and elderly people she visits in her church volunteer work.

Their greenhouse has become something of a conversation piece in the San Luis Valley because the Vigils have been able to grow coriander, which is highly prized by many valley residents as flavoring for regional dishes. Teresa also provides garden starts to neighbors and friends and sells catnip grown in her greenhouse to a local pharmacy.

In addition to food, the Vigils' greenhouse provides supplementary heat to their house. Because it was attached to the coldest part of the house, the heat it provides to their south bedroom is most welcome. Even though floor space in the greenhouse is limited, the Vigils' seven grandchildren naturally gravitate toward the greenhouse when seeking play space. Victor and Teresa are planning to build a new solarium/playroom that will, of course, be solar heated and provide additional heat to their kitchen. They will build the new room with logs, rocks, adobe, or whatever else they can find near them that is either free or very inexpensive. Teresa explained: "I believe you should build with whatever is around you."

*1-32. Teresa Vigil with her towering tomato plants (Photo: Darryl J. Strickler)*

# Rita and Dick Scarlett
# Grass Valley, California

Although Rita and Dick Scarlett are both senior citizens, you could hardly refer to them as retired people. Rita has an active private nursing practice, and Dick spends his days in a variety of pursuits. The geodesic dome house they live in was designed by Rita in 1977. Shortly after their home was completed, the Scarletts began to read books and articles on solar heating with the thought of adding a greenhouse to their home. They realized that adding a greenhouse would not be a simple task, and of course, no predesigned examples of how to attach a greenhouse to a dome structure existed. The job called for someone who had expertise and experience in the area of solar design, as well as creativity and skill in engineering. The Scarletts located just such a person: David Norton, of Habitat Constructions, who had recently moved to the Sierra Nevada foothills from Davis, California, where he had been associated with the Village Homes all-solar subdivision.

The Scarletts told Norton what they wanted, and thanks to Norton's skill as a designer/builder, they now have exactly what they asked for: a working passive solar greenhouse in which they grow houseplants, tomatoes, lettuce, broccoli, onions, and parsley.

*1–35. Interior of Scarlett greenhouse showing heat storage system (Photo: David Norton, Habitat Constructions, Grass Valley, CA)*

All of the food is grown in raised beds with wheels on them that Norton designed and built.

The Scarletts' solarspace was built on an existing deck to which additional structural support was added. It was completed in 1979 at a cost of approximately $6,000.00 (about $37.50 per square foot). This price included materials and labor to reinforce the deck, the growing beds, and thermal storage system. The Scarletts' solarspace has a trapezoid-shaped floor that is about ten feet wide and sixteen feet in length along the south wall. The east and west walls have removable, insulated wall sections that are taken out in summer to allow flow-through ventilation. The rafter extensions over the sloped south glazing of the solarspace hold a canvas awning that is put in place during the summer to shade the glazing. The floor of the solarspace has a three-inch-thick concrete slab poured over plywood and insulated on the underside. Additional thermal storage is provided by five-gallon containers stacked on specially designed shelves along the north wall of the solarspace.

The Scarletts heat their home primarily with wood. The supplementary solar heat from the greenhouse enters the house through a sliding glass door. Airflow is by natural means only; no fans are used.

Rita and Dick live in a unique house that has a one-of-a-kind solarspace attached to it. With the help of David Norton, they have provided all present and would-be dome dwellers with a first-rate example of how a solarspace can be integrated into dome architecture. In addition, they have provided a glimpse of a future based on renewable energy resources and home-based food production.

*1–34. Uniquely designed greenhouse on Scarlett dome residence (Photo: David Norton, Habitat Constructions, Grass Valley, CA)*

# Diane and Gary Meier
# Shawnee Mission, Kansas

Diane and Gary Meier's solarspace evolved from their well-developed interest in conservation and renewable energy sources. Diane's advocacy of safe energy dates back to the early 1970s, when she served as a lobbyist and volunteer, and continues to the present in her role as president and cofounder of Energy for Rural Self-Reliance, a nonprofit education and consulting group.

The need to gain firsthand experience and produce a working example of solar energy for home heating strongly influenced the Meiers' decision to add a solarspace to their home in 1979. At the time the potential for food production was not a major priority; both worked full time and were actively involved in raising their children, Kristen and Walter. They felt they did not need something else to do.

Since they built the solarspace, however, the Meiers' priorities have changed somewhat. Their interest in growing plants began somewhat tentatively at first but over the years has developed into a fulfilling hobby. They use the solarspace to start plants for their outdoor gar-

den and as a year-round residence for a variety of houseplants. Diane now finds that working with plants has its own special rewards.

To be sure, the Meiers are still interested in the 25 to 30 percent savings on their home heating costs, but they have also discovered many other benefits their solarspace provides. In keeping with one of their original goals—to provide an example of solar energy use—the Meier solarspace has been visited by literally hundreds of people on tours, including entire classes from Kristen's and Walter's schools. The solarspace has also become the focal point of their home—so much so that, after building the solarspace, they rearranged the interior of their home to be more visually and spatially connected to the solarspace. When guests arrive at the Meier residence, they go directly to the solarspace; in fact, it is difficult to get them to sit anywhere else.

The eleven- by twenty-two-foot solarspace was designed and partially completed by Dale A. Clark, a solar designer/builder. It was built on a pier foundation because the access to the backyard was too limited to allow for a poured concrete foundation. Reflecting the Meiers' conservation ethic is the framing lumber for their solarspace, recycled from a redwood deck that

*1-36. Dining area of the Meiers' solarspace (Photo: Bruce T. Novell)*

*1-37. Exterior of Meier solarspace; note removable glazing panels (Photo: Bruce T. Novell)*

formerly occupied the spot where the solarspace now stands. The finish work and other refinements were done by the Meiers and contractor Greg Des Marteau, who now specializes in solar construction and remodeling. The total cost of the project was $5,500.00 ($22.73 per square foot in 1979).

A unique feature of Meiers' solarspace is the glazing units in the south wall that can be removed and replaced by screens in summer. Designing and building a removable glazing system that will not leak air or water are certainly not easy tasks, nor is it easy to remove and replace the heavy double-insulated units in the spring and fall. The Meiers feel, however, that the connection between the solarspace and the outdoors is greatly improved by the screens in summer, a factor that is very important to them.

Equally important to them is the biological connection between their house and solarspace. Thus far, they have used only natural means to control unwanted insects in the solarspace. Because they are understandably reluctant to use chemical pesticides, they have imported five chameleons and a garter snake or two to check the insect population. Before the chameleons escaped to the outdoors, they each ate a healthy share of flying insects and a few sowbugs and, of equal importance, provided a dazzling display of changing color and movement in the solarspace.

## Leon and Roberta Brauner
## Bloomington, Indiana

Many people build a solarspace to provide additional heat to their home or as a place to grow plants. Leon and Roberta Brauner had another important reason for building one: they needed to expand their nursery school. Rather than build a more traditional addition onto their house, they decided to design a structure that would provide the majority of its own heat during the winter and stay cool during the warmer weather. When they began planning the addition in 1979, however, not many books or people were available to provide the kind of specific information the Brauners needed. This fact hardly slowed them down at all; they read magazine articles, library books, and whatever else they could find and sought the advice of anyone they thought was knowledgeable. Leon, a member of the Indiana University theater design faculty, produced many drawings and cardboard models before he arrived at a solution he thought would work.

With the help of craftsman-builder Larry Doll, the Brauners began constructing their solarspace in August, 1980. The excessive amount of native limestone in the area where the solarspace was to be constructed made the foundation work very difficult, but the project picked up pace as the framing took shape. Doll worked during the daytime, and the Brauners and their children, Michael, Craig, Christine, and Mark, joined the effort after school and on weekends. Throughout the construction special attention was given to building the solarspace ''tightly.''

The solarspace was completed in November, 1980, and was immediately pressed into service for its intended purpose: the daytime headquarters of a group of lively three- and four-year-olds and their teachers. The children who attend Roberta's Small World School thoroughly enjoy being in the solarspace. They naturally gravitate toward the more sunny spots like any good solar kid would. Brauners' nursery school provides an excellent working model of solar heating and cooling, not only to the children who inhabit it on a regular basis, but also to their parents, who seem to appreciate the open, natural environment.

Leon and Roberta's solarspace cost

1-38. The Brauners' solarspace/nursery school (Photo: Darryl J. Strickler)

$13,500.00 ($22.00 per square foot). Federal and state tax credits—which, in retrospect, Leon says he figured conservatively—reduced the net cost of the project to $18.83 per square foot. Levolor miniblinds are used to control sunlight and reduce glare and are also tightly closed on winter nights to serve as a radiant barrier to reduce heat loss. In addition, rigid insulating panels are placed over the south glazing of the solarspace on winter nights. Existing windows in the south wall of Brauners' house are opened during sunny winter days to admit solar-heated air. The clerestory windows in the roof of the solarspace provide both light and heat and, when opened in the summer, enable hot air to escape from the solarspace.

Throughout the hot and humid Indiana summers, the combined effects of a good ventilation scheme and deciduous trees to the south keep the solarspace cool and comfortable. Because the solarspace/nursery school now covers the previously exposed south-facing masonry wall of Brauners' house, the interior of the house also remains cooler during the summer.

1-39. Nursery school in progress (Photo: Leon Brauner)

## Barbara Samuel and Tom Lamm
## Madison, Wisconsin

Tom Lamm and Barbara Samuel's decision to build a passive solar addition on their home in 1980 was definitely not made in haste. For the previous ten years, they had been actively involved in reducing their dependence on non-renewable energy and centralized food supplies. During the early 1970s when they lived on a small farm near Madison, Wisconsin, they gained a great deal of practical experience in organic gardening and small animal husbandry and were successful in developing a life-style based on energy and resource conservation and home-based food production. As energy prices began to climb after the 1973–74 Arab oil embargo, however, they found it increasingly difficult to justify the amount of gasoline they were using to get to their jobs in Madison. (Barbara is the coordinator of public education and information at the Wisconsin State Energy Office; Tom is on the staff of the University of Wisconsin-Extension Environmental Awareness Center.) Also, their growing concern over the dual trends toward natural resource depletion and urban sprawl made them curious to see if self-sufficiency could be attained in an urban environment. So, in 1977, they bought a home in the city.

Barbara and Tom were happy to discover that the fifty-year-old house they purchased in Madison already had storm doors and windows and some insulation in the attic. Over the next three years, they completed the job of insulating and weatherizing by caulking and weather-stripping doors and windows, insulating the basement ceiling and sillboxes, and adding attic insulation.

With a well-sealed and insulated house, Tom and Barbara turned their attention to the addition of passive solar features. They had earlier decided to add space to their long, narrow living room, and they felt that making it a solarspace would bring multiple benefits. Barbara first contacted Donald Schramm and Paul Luther of PRADO, a solar design and consulting firm in Madison, in 1979. Because of previous commitments, however, PRADO was not able to begin the design work for nearly a year. As it turned out, the wait and the six months of planning before the construction actually began,

paid off very well because it gave Tom and Barbara time to fully explore their ideas and, later, to "live" with the design. They now have an attractive and well integrated solarspace that enhances not only their house, but also their lives.

*1–40. The Lamm/Samuel two-story solarspace (Photo: Donald Schramm, PRADO)*

Although it is probably difficult for anyone who has not visited Barbara and Tom's house to imagine how a two-story addition, only 4 feet deep and 15 feet wide (about 120 square feet), could make any difference in how they live and perceive their relationship to nature, Tom and Barbara's solarspace has done just that. It provides substantial light and warmth to the interior of their home and has helped them become more energy self-reliant.

The remodeling design included the addition of an airtight wood-burning stove capable of heating the entire 1,400-square-foot house. Because Barbara and Tom firmly believe that urban wood users should use urban wood supplies—rather than using nonrenewable energy to search for and transport firewood from rural areas—they always keep an eye out for nearby sources of free firewood. Their stable wood supply is from oak pallets discarded by a local transporting operation, but they regularly pick up scrap lumber left on neighborhood curbs for the trash collector.

Since their solarspace was completed in the spring of 1981, Tom and Barbara have not used their central heating system. (They light the pilot of their gas furnance only when they plan to be away for extended periods of winter.) During the first winter, they were surprised and delighted to discover that wood heat is more reliable, comfortable, and immediately responsive than other heat sources. They were happier still when their monthly winter gas bills stayed under twenty dollars, whereas those of friends and neighbors soared into the hundreds.

The solarspace is ventilated in summer by opening the casement windows on the east and west sides. In winter insulating curtains are used at night and on cloudy days to reduce heat loss through the glazing. Barbara and Tom spent a total of $14,000.00 on their solarspace and feel it was money well spent. Combined federal tax credits and state incentives reduced their total net cost to $11,000.00 ($91.66 per square foot of floor space).

1-41. A closer look (Photo: Donald Schramm, PRADO)

# TWO
# Choosing Your Solarspace

What kind of solarspace will best fit your needs and the structure of your house?

Now that you have read about other people's solarspaces, you are probably ready to begin thinking about your own. The questions you must answer first are: how will the space be used, by whom, at what times of the day, and in which seasons of the year? The answers to these questions will help you decide what kind of solarspace to build. You can begin to form your answers by considering each of the major functions a solarspace can serve.

## PRIMARY FUNCTIONS

The three primary functions of a solarspace are to supply supplemental heat to the house; to add floor space; and to provide a place for growing food and/or houseplants. Although these uses are not mutually exclusive, as demonstrated by the examples in chapter 1, all three can be served by the same solar structure, be it a greenhouse, sunspace, solarium, or enclosed porch. You will probably discover as you read further, however, that you will be more satisfied with your completed solarspace if you select one of the three primary functions as your *major* emphasis.

In chapter 3 you will also find that the primary function you select should influence the actual design of your solarspace. If you consider each of the major functions separately, you will see how they are interrelated and will be able to decide more easily how your solarspace will be used, by whom, and when.

## Supplementary Heat

Many people who have decided to build a solarspace are primarily interested in energy savings. In most climates a properly designed solarspace can, in fact, supply all of the heat required for a well-insulated, weather-tight house or building on a sunny winter day. A space that provides this much heat can be designed into the plans for a new house, or it can be added to an existing house. Unfortunately, not all existing houses can retain the heat produced by a solarspace—or a furnace for that matter; it leaks out through windows, walls, ceilings, floors, and cracks around doors and windows. Your first concern, therefore, must be to make sure your house is adequately insulated and sealed.

For the majority of existing houses, including those in the examples in chapter 1, it is most realistic to consider a solarspace as a source of *supplemental* heat for the house. In fact, a solarspace literally provides *only* supplemental heat overall, since it produces virtually *no* heat on an extremely cloudy day. Depending on the climate and the actual size and configuration of a solarspace, it can potentially provide anywhere from 10 to 80 percent of the yearly total of Btu's required for home heating.

The amount of thermal mass a solarspace has in its interior will affect the amount of heat it can provide to the house. Thermal mass can be added to a solarspace by using concrete, brick, stone, adobe or other masonry materials as structural elements; or it can be added by placing containers of water or phase change materials inside the solarspace. Adding thermal mass to a solarspace enables it to store heat produced on sunny days during the heating season, thus helping to keep the space warm at night and during cloudy weather and reducing daily temperature fluctuations within the space. The solarspace will therefore usually be comfortable for people or plants that inhabit it.

A solarspace with no thermal mass—other than that contained in the low-mass materials, such as lumber, used to build the space—can climb to more than 100°F (38°C) on a sunny winter day and drop to nearly the outdoor temperature the same night. If the existing south wall of your house has little mass—for example, if it is of typical light-frame wood construction—you may find it more cost-effective to use your solarspace primarily to supply heat to your house during the daytime on sunny days during the heating season (see further discussion of these contraints in the following section, "Your House").

The specific amount of thermal mass a solarspace should have must be related to the function of the space (see chapter 3). If the space will be used for food production, living-related functions, or both, it must include enough thermal mass to keep it within a tolerable range of comfort for people and plants—usually 50° to 85°F

(10° to 29°C). On the other hand, if the solarspace is to be used solely for supplementary heat production during the heating season and the comfort of people and survival of plants is not a major concern, additional thermal mass in a solarspace is unnecessary. To effectively use the excessive heat produced by a low-mass greenhouse, sunspace, solarium, or enclosed porch, the heat must be drawn off with fans and directed into the house, into the supply side of a forced-air furnance, or into a thermal storage bed located under the house (see chapter 3).

The time at which a solarspace provides heat is an important consideration as you begin to think about adding such a space to your house. If the orientation of a solarspace is close to true south (the direction of the sun at solar noon), it will reach its maximum temperature on a sunny day around 1:00 P.M. and its minimum around dawn. If the following day is clear and bright (or even hazy and bright), the space will begin to heat up again. If the day is overcast or rainy and cold, the space will continue to lose heat until it levels off at its minimum temperature. As the temperature in the solarspace drops below the temperature in the house, the house begins to lose heat to the solarspace. The process reverses itself when the solarspace is warmer than the house: the solarspace loses heat to the house and, as always, to the outdoors. The greater the temperature difference between the outside air and the air in the solarspace, the greater is the heat loss.

All of this may be a bit easier to understand through an example that illustrates how to "operate" a solarspace. Suppose you build a sunspace on the south side of your house; how do you regulate the heat produced? When the sunspace is warmer than the house during the heating season, you simply open the doors, windows, or vents between the house and the sunspace. The heated air in the sunspace will move into the house, and the cooler air from the house will return to the sunspace by natural convection. Depending on the size of the solarspace and the rate of airflow, the total volume of heated air in the solarspace will be exchanged every two to five minutes. To block off this exchange of air when the sunspace is cooler than the house, you simply close the doors, windows, and vents to the sunspace.

A solarspace will typically lose a great deal

of heat through its large glazing surface, especially during clear or very windy nights in winter. Depending on your climate, it may be cost-effective to install insulating or radiant devices that can be placed over glazing surfaces at night and during cloudy weather to reduce heat loss from the solarspace. Insulating drapes, rolldown shades, shutters, louvers, rigid insulation board, and other materials can be used to slow down the flow of heat through glazing surfaces when the sun is not shining (see chapter 6 and the Appendix).

If no one will be home at the appropriate times to regulate the openings between the solarspace and the house, you can install automatic vent operators or thermostatically controlled fans to move the solar-heated air into the house. The air-distribution system can be as simple or complex as you care to make it. For example, if you have a central forced-air heating system, you can duct air from the solarspace to the furnace so it can be distributed throughout the house. (If you use this strategy, include a backflow damper on the system to prevent circulation of cooler air from the solarspace at night.)

If you plan to use natural convection to move air in and out of the solarspace, check the existing furnace outlets and return ducts in rooms adjacent to the solarspace to determine whether these outlets can be closed when the rooms are heated by the solarspace. Also check to see whether the solar-heated air would affect the existing thermostat. If hot air from the solarspace passes over the thermostat, it could influence comfort levels in other parts of the house that do not receive solar-heated air.

The air from a solarspace that contains many plants has a higher moisture content than conventionally heated air because the plants give off moisture. This helps to combat the dry air conditions found in most homes during the heating season and may help reduce cold-related illnesses of the occupants of the house. Moreover, the higher the moisture content of the air, the warmer it feels. Therefore, the quality and thermal comfort provided by air heated in a solarspace is generally superior to air heated, for example, in a furnace. As an added benefit, plants in a solarspace remove carbon dioxide from the air and release oxygen. In effect, plants "scrub" the air entering the solarspace from the

2-1. Window Quilt Insulating Panel (Photo: courtesy of Appropriate Technology Corporation, Brattleboro, VT)

house and return oxygen-rich air to the house—something no furnace can do.

After you have paid back the cost of building your solarspace through savings resulting from decreased use of utility-supplied energy, the heat you produce in your solarspace is essentially free of charge. As the cost of energy from conventional sources increases over time, the payback period is shortened and the value of the solarspace in terms of dollar savings increases. Furthermore, most solarspaces require very little maintenance to keep them functioning for heat-production purposes. Basically, only regular cleaning of the glazing areas during the heating season and occasional recaulking are required. This means that the actual yearly expense for maintenance—other than your own time—is limited to the cost of a gallon of window-washing solvent and a couple of tubes of caulk. (Although it will probably not affect the heat-producing capabilities, you may also want to do some periodic repainting or refinishing of the structural elements of the solarspace.)

## Additional Space

Strictly speaking, the floor space in a greenhouse, sunspace, solarium, or enclosed porch should not be considered living space. Because of temperature fluctuations, the added space should be viewed in terms of specialized uses during specific times of day and specific weather conditions. For example, you probably would not want to spend much time in sedentary activities in a solarspace that had dropped to its minimum temperature. But, if the temperature in the space had dropped to, say, 58°F (14°C) overnight, you may find it a great place to do your morning exercises—especially when you consider that the plants would appreciate the increased carbon dioxide you give off through huffing and puffing.

On a cold winter day, you might appreciate putting on a pair of shorts and sitting in a solarspace that had reached 85°F (29°C) to write a letter to your Aunt Harriet, but you would probably not have the same inclination during warmer weather. Most well-designed solarspaces are habitable and enjoyable places to spend a sunny, cold day, but when the sun goes down and the space begins to lose heat, you will probably want to head back into the house—especially if you do not have nighttime insulation for the glazing. When warmer weather comes along, you will no doubt prefer to spend your time outdoors. Some people and solarspaces provide the exceptions to the above generalizations, but for the most part they hold true.

The question is not so much one of how to use the additional space provided by an attached solar structure—unless you have already decided to build one solely for its heat-producing potential. If you need to add space to your house, the question should be what kind of space will fill that need. For instance, you may be more satisfied with a typical family-room addition—with lots of south-facing windows for direct solar gain or with a thermal storage wall—than you would be with the types of solarspaces described in this book. A family room could be designed so that it would be solar heated but

2-2. A solarspace can provide a happy home for your houseplants (Photo: Darryl J. Strickler)

would probably not supply much heat to the rest of the house. Certainly if what you need is additional bedrooms, for example, a solarspace will not fill this need, although a solar addition that includes bedrooms might.

Solarspaces can be used for many purposes in addition to food and heat production. These include using the space as a play area for children; a mudroom and air lock entry to the house;

2-3. Solar clothes dryer (Photo: Darryl J. Strickler)

a "solar clothes dryer" (which is an excellent way to add moisture to your house in the winter); an enclosure for a hot tub or spa; a non-freezing environment in which to install a pre-heating system for domestic hot water; an additional casual dining area; and a workshop, studio, home gym, or gameroom.

Many solarspaces seem to have mystical qualities that can not easily be explained in terms of their structural elements. These qualities seem to have more to do with the interplay of light and shadow along with the living vibrations of plants—all of which exist in a medium of air that has a special essence rich with nature. When you have a solarspace full of plants connected to your house, the house and the solarspace develop a unique relationship with one another—one that can potentially improve the quality of life of the inhabitants of both the solarspace and the house.

2-4. *Plants plus light equal magic (Photo: Paul Gallimore, Long Branch Environmental Education Center, Leicester, NC)*

## Space for Growing

Most existing solarspaces have at least a few houseplants in them, but less than half of all solarspace owners seriously attempt to grow food in them. Perhaps one of the reasons for this is that few people have the experience or want to take the time required to grow food in their own home. Furthermore, many people simply find it

2-5. *Greenhouse harvest (Photo: David Strickler, Strix Pix)*

easier to buy what they need rather than grow it themselves. This trend will probably not change rapidly as long as we are able to purchase relatively fresh produce at a reasonable price. What may seem a reasonable price to one person may, however, seem outrageous to another, and what may seem to be a moderate increase today may one day become intolerable.

The matter of food prices and availability has many important implications as you begin to consider adding a solarspace to your home. The most obvious of these is that, no matter what the primary function of your solarspace is, it should be planned so that you *could* grow food in it—even though growing food may not seem so important at the present time. Leave some open spaces in the floor area to serve as planting beds or plan to use shelves or racks for potted plants or hydroponic troughs. Similarly, if you plan to build a greenhouse primarily for the purpose of horticulture and/or aquaculture, you should allocate some floor area for other functions in addition to food production.

What kind of food can be grown in a solarspace? Such traditional cold-weather vegetables as lettuce, spinach, cabbage, brussels sprouts, and oriental vegetables can be grown during the winter in most climates. Herbs and some fruits can also be grown if the solarspace is properly designed to provide adequate light levels and remains within a tolerable temperature range —generally, 50° to 85°F (10° to 29°C). In addition, some varieties of fish such as tilapia, a tropical fish that provides high-quality protein, will thrive and grow in solar-heated tanks inside a greenhouse, sunspace, or solarium. What you grow in your solarspace should be related to what you like to eat and to what ''crops'' will grow best in your climate—both the outdoor climate and the indoor climate you create in your solarspace. Chapter 6 contains a more detailed discussion of food production.

While you are considering what you could grow, you should also think about how much time will be required to manage food production in your solarspace. Obviously, the more food you grow, the more time and effort it takes. For example, if you have a fully planted 8- by 24-foot (2.4- by 7.3-m) attached solar greenhouse that includes growing beds, racks, hanging plants, and fish tanks, you would need to spend several days a year changing the plants and soil in the planting beds, repotting plants, and fertilizing the soil. Daily, you might spend thirty to forty-five minutes watering plants, pulling weeds, and controlling insects. Yes, insects—virtually nothing can keep them out. In fact, you may not want to keep out such things as ladybugs, praying mantises, and bees (see chapter 6). For now, consider it part of the deal—if you are going to grow a lot of food, you may have at least aphids and whiteflies, both of which are relatively easy to keep in check.

Even this preliminary survey of food production would not be complete without some mention of the major reasons for growing things in the first place: the satisfaction of becoming an integral part of the growing process, making a real connection with nature, and receiving the fruits of your labor in the form of visual splendor or food. These are certainly among our most basic human needs. Although they seem to have been long forgotten in our evolutionary process, they are beginning to be rediscovered by many people. Creating an ecosphere in your own home will allow you and the people you live with to join in this rediscovery process.

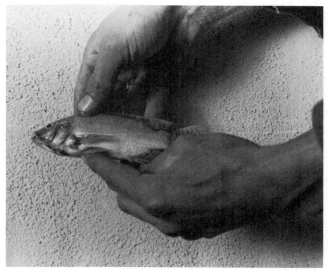

2-6. *Tilapia* (Tilapia aurea) *or blue tilapia (Photo: Arnie and Maria Valdez, People's Alternative Energy Service)*

2-7. *James Cason raises tomatoes in his attached solar greenhouse (Designer-builders: Portland Sun; photo: Darryl J. Strickler)*

# YOUR HOUSE

Having considered the basic functions that can be served by a solarspace, you should be closer to a decision on how you want to use your solarspace. The question of what kind of solar structure to build must be related not only to the needs of the people in your household, but also to the structural elements and orientation of your house. As you continue to explore the possibility of building a solarspace, consider these important questions:

■ What building codes, restrictions, covenants, and easements apply to your property?
■ What is the exact orientation of your southernmost wall toward true south, the direction of the sun at solar noon?
■ What is inside, behind, outside, and beyond the southernmost wall of your house?
■ How thermally efficient is your house?

2-8. *Check your setback restriction and solar access rights (Photo: Darryl J. Strickler)*

## Building Codes, Restrictions, Covenants, and Easements

Before you proceed any further, check any legal restrictions that apply to your property or to any improvements you wish to build. Building codes vary from state to state and city to city. For example, your property may be subject to a setback restriction that prohibits you from constructing anything closer than a given distance from your property line or the street, or a specified coverage limit may prohibit you from building on more than a given percentage of your lot. Your property may also be subject to local building codes that specify the type of foundation or the type and amount of glass you may use for your solarspace. Begin by calling the local government agency that is responsible for building code enforcement.

While you are in the process of checking codes and restrictions, you can also find out if a building permit is required and how to apply for one. Also find out if your state or local government has enacted legislation covering solar access rights. Without such laws you have no guarantee that someone who owns the adjoining property to the south, east or west would not

build or grow something that could prevent the sun from striking your solarspace. If you do not own enough land to ensure solar access, find out whether your particular development or subdivision could develop a "neighborhood restrictive agreement" that covers solar access or try to arrange a long-term legal agreement with the owners of property that adjoins yours.

## Orientation

Since the orientation of the south wall of your house will influence the amount of solar radiation that strikes the glazing of a solarspace attached to the wall, it is important to know the exact orientation of your south wall. The maximum amount of solar radiation strikes a wall that faces true south—180 degrees adjusted for magnetic variation. A surface that faces 30 degrees east or west of true south will collect about 90 percent of the amount it would collect if it faced true south. If you are planning to include a solarspace in a new house, you can orient the south wall of the house toward true south or use any variation from true south that you prefer. Unfortunately, if you are adding a solarspace to an existing house, you will not be able to do much about the orientation of the south wall.

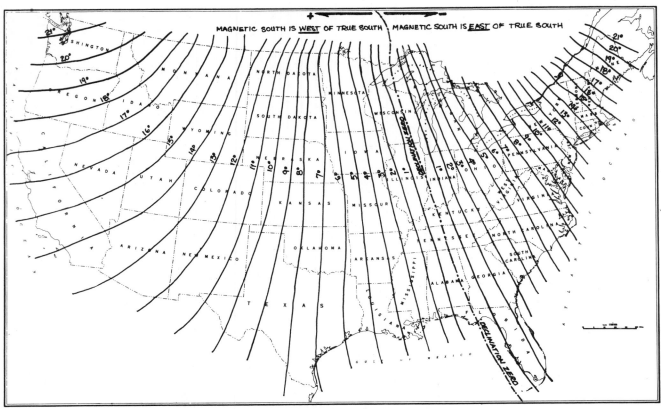

2-9. Compass variations from true south (Based on map #1-1283: Magnetic Declinations in the United States—1980; Department of the Interior, U.S. Geological Survey, 1980)

To determine the orientation of your southernmost wall, use a compass with a rectangular base on which the dial can be rotated so the compass can be accurately aligned to the wall. Remember that the compass will indicate the orientation of your house toward magnetic south, which can vary from a few degrees from true south to as much as 22 degrees, depending on the magnetic variation (declination) of your locality. In 1980 the line of zero declination (ZD) stretched roughly from the western shore of Lake Michigan through the center of the Florida panhandle (see fig. 2–9). If your house is located near the ZD line, the reading on a compass will indicate true direction. If you are *west* of the ZD line, magnetic south is *west* of true south, so you must *add* the number of degrees of declination for your locale to the compass reading to find the orientation of your south wall. If you live *east* of the ZD line, magnetic south is *east* of true south, so you must *subtract* the number of degrees of declination.

If reading a compass and making adjustments for magnetic declination is not within your capabilities (or tolerance level), try looking at the shadows cast by the sun coming through a window or a door in your southernmost wall at about 12:00 noon standard time (1:00 P.M. daylight time). If the shadows cast by the vertical elements of the window or door frame are perpendicular or nearly perpendicular to the wall surface—that is, if the sun is coming "straight in"—your south wall is oriented close enough to true south to build an attached solarspace on the wall. (The actual time of solar noon, when the sun is at true south, is midway between the time of sunrise and sunset.)

In some climates having the south wall of a solarspace oriented to the east or west of south is actually more desirable. A solarspace oriented toward the south-southeast will heat up earlier in the morning in winter and will face away from the hot afternoon sun in summer. In areas subject to early morning fog, such as coastal areas, a solarspace that faces slightly west would potentially perform better since it could collect the sunlight later into the afternoon after the sky had cleared. A solarspace that faces more than 15 degrees west of true south should have some east wall glazing, especially if you plan to grow plants in it. If you are a late riser, you might prefer a solarspace that faces west of south, since the early-morning warmth and light from the sun would not be as important to you. Similarly, if you are an early riser, you may want your solarspace to face east of south so you can receive the most benefit from the early sun.

To receive an adequate amount of solar radiation, the south wall of your house (and the solarspace you attach to it) should face no more than 30 degrees east or west of true south—assuming that you want the south wall of the solarspace to be parallel to the south wall of the house. If your southernmost wall is more than 30 degrees from true south, you might consider building a solarspace on the corner of your house so that the south glazing of the solarspace faces true south (see fig. 2–10).

There is one possible exception to the 30-degree limit. If you live in a very cloudy climate, such as the Pacific Northwest, and if you plan to use roof glazing in your solarspace to collect diffuse sky radiation associated with cloudy conditions, the orientation of the south wall of the solarspace is less crucial, since a majority of radiation will be collected through the roof area.

## The South Wall

After you have determined the orientation of your southernmost wall, you should consider what is behind, beyond, inside and outside that wall.

### Behind

What is behind the wall refers to the rooms or living spaces adjacent to the south wall. Are they bedrooms, living or dining areas, a kitchen? Or are they utility areas such as a bathroom, laundry, or storage closets? It is usually most feasible to add a sunspace or solarium to living, dining, or kitchen areas—especially if the solarspace is to have multiple uses that include dining, working, or just relaxing when the weather permits. Here physical and visual connections to the living areas are important to allow people and heat to move freely between the solarspace and living/dining areas of the house. On the other hand, a solar greenhouse used primarily for horticulture or a low-mass solarspace intended mostly for heat production could work very well if it were attached to utility areas of the home, even a garage, for example. Since a greenhouse or low-mass solarspace would presumably be used less often for living-related functions and would probably not be finished out as attractively, its connection to living areas would not be as important.

### Beyond

What lies beyond your south wall? How far is your property line (or the line required by setback restrictions) from the wall? Is there significant shading of the wall in winter by evergreen trees, shrubs, or the trunks and bare branches (or dead leaves) of deciduous trees? Is the south wall shaded by a tall building or a neighbor's house? Is the south side of your house visible to neighbors or passersby?

The most ideal situation for adding a solarspace would be: where you have adequate distance between your house and the property or setback line, unobstructed access to the winter sun and shade from the summer sun provided by tall deciduous trees, and enough privacy afforded by the orientation of the house, or fences and shrubs, so that neighbors and people going by could not see into your solarspace. The last point is obviously a matter of personal preference, since you might want people to be able to see your solarspace. What matters most is that the *sun* be able to see your solarspace, at least from 9:00 A.M. to 3:00 P.M. during the heating season, and that you have enough room on your property to build a solarspace. (A minimum of 10 feet (3m) between the house and the property or setback line is usually adequate.)

### Inside and Outside

Finally, you need to consider how and of what materials your south wall is constructed. The most ideal wall on which to construct a solarspace is a solid masonry or adobe wall, 8 to 12 inches (20 to 30 cm) thick, with no insulation or air space in it. Such a wall will store heat collected in the solarspace and will allow the heat to pass through the wall into the house. Unfortunately, in addition to being ideal, this type of

2-10. *Solarspace facing south on a house that does not (Photo: Paul Gallimore, Long Branch Environmental Education Center, Leicester, NC)*

wall is also very uncommon. The majority of houses built in the past thirty years have stud-framed walls with insulation between the studs and siding or masonry veneer as the outside covering. Brick or stone on the outside of the wall will store heat in a solarspace attached to it, thereby reducing the temperature fluctuations in the solarspace and making it more comfortable for people or plants that inhabit it. If the wall behind the exterior veneer is insulated, however, the heat produced in the solarspace cannot travel through the wall to reach the house. If you plan to use your solarspace primarily for supplementary heating, that heat must be extracted by natural convection or with fans.

At this point you should consider how much of the required thermal mass for your solarspace is already in (or could be added to) your south wall. What do you do if your south wall is of low-mass, light-frame construction? Two basic approaches can be used to solve this apparent problem: add mass to the wall or to the interior of the solarspace if people or plants are going to inhabit the space in winter; or use the solarspace primarily to provide heat and use duct fans to draw off and recirculate the excess heat that is characteristic of low-mass solarspaces. If your solarspace will be attached to a frame wall, keep these approaches in mind as you begin to plan it.

## Thermal Efficiency of Your House

As suggested earlier, it would indeed be foolish to add a solarspace to produce heat if that heat immediately leaked out of your house. Given the nature of heat flow in a building, you can count on heat to travel through windows, walls, floors, ceilings, and cracks around doors, windows, and vents to reach the cooler outdoor air. You can also count on air from outdoors finding its way into your house. The basic goal is to slow down the heat flow through exterior surfaces by setting up resistance in the building's envelope. The designation $R$ is an indication of the resistance of various materials or surfaces to heat flow. The higher the R value of a surface, the greater its resistance.

Your house must be adequately insulated and sealed against air leaks before you consider adding a solarspace to it. Adequate insulation varies with the climate of the region you live in. In cool climates a minimum of R-19 walls, R-30 ceilings, and R-10 perimeter insulation would generally be considered adequate. Slightly less insulation is required in warmer climates; more is desirable in colder climates. If your house does not meet these standards—and surprisingly few houses do—you should definitely upgrade it if possible. At the very least, you must seal all the cracks with weather stripping and caulking. Infiltration of air through cracks and opened exterior doors causes 40 percent of the winter heat loss and summer heat gain in most houses. In addition to sealing cracks, you might consider replacing badly weathered windows, adding storm windows and doors, and building an air lock vestibule for entry doors, or better yet, use your solarspace as an air lock entry. All of these improvements will reduce your heating and cooling bills and allow you to make better use of solar-heated air.

If you need additional information on insulating and weatherizing your house, check the excellent sources listed in the Bibliography. Do not get the idea that the brief treatment of thermal efficiency above indicates that weatherization and insulation are of little importance. In fact, they are vitally important, so get to it! After you have a thoroughly weatherized and insulated house, you can move to the question of how to design your solarspace.

---

### Radon

Radon is a naturally occurring radioactive gas produced by the decay of radium 226, which is a by-product of decaying uranium 238. Chances are good that you already have traces of radon around and *inside* your house because radon is normally present in soil, natural gas, water, and in many building materials.

Normally radon poses no significant danger to humans because it can readily escape to the atmosphere. A potential health hazard occurs, however, when a tightly enclosed space, such as a house or solarspace, is built over a small portion of the earth and additional masonry such as concrete, stone, or brick, which also contain radon, is added. If all the cracks and joints in the structure are carefully filled, the radon cannot easily escape, especially when the structure is tightly closed for extended periods, as during the winter months.

At present the effects of radon and the specific amount that poses a threat to human health are the subject of much debate and inquiry. Awareness of the possible effects of radon should certainly not prevent you from building a well-sealed house or solarspace. You should, however, understand that good ventilation will greatly reduce any potential dangers of radon . . . and a whole host of other sources of indoor air pollution as well. Although opening a window to admit fresh air during the winter months (more accurately, to allow heated air to escape) is not usually practical, it is certainly practical to install an air-to-air heat exchanger in your home, be it a whole-house unit that is tied into the existing duct work or window units placed strategically around the house or in the solarspace. See chapter 3 and the Appendix for more information on air-to-air heat exchangers and list of their manufacturers.

# THREE
# Designing
# Your Solarspace

What size and shape solarspace would best serve your needs and the characteristics of your house? How much will it cost you to build? How much glazing and thermal mass should it have? How should you plan your solarspace?

Any type of planning involves problem solving. The problem to be solved in this case is how to achieve the most cost-effective, energy-efficient, aesthetically pleasing solarspace, given your goals, your financial situation, your do-it-yourself skills, and the characteristics of your house. This chapter will help you make the initial decisions and guide you through the steps necessary to complete the planning of your solarspace. Chapter 4 will help you decide what assistance you need to complete the design and construction of your solarspace, and chapter 5 will tell you how it should be built.

## RELATING FORM TO FUNCTION

The characteristics of your house and the primary function you designate for your solarspace should influence the size and shape of your solarspace and the materials used to construct it. In other words, the design of your solarspace should be related to the purposes for which you plan to use it. The space should also be designed to harmonize with the overall style of your house so that it will appear to be integrated rather than tacked on. In short, your goal should be to plan a solarspace that will increase the value of your property, not depreciate it.

If you begin by considering some general guidelines, you should be able to make the formative decisions that will affect the final design of your solarspace. Although the guidelines below contain very general statements, the information in each generalization is explained in greater detail later in the chapter.

## A Dozen Guidelines for Designing a Solarspace

1. An attached solar greenhouse intended primarily for food production should have sloped south-facing glazing, as well as some glazing in the east and/or west walls and in the roof. If a greenhouse is at least twice as long as it is wide and it faces close to true south, east- and west-wall glazing may be

unnecessary. Regardless of its length-width ratio, however, a greenhouse that faces more than 15 degrees west of true south (195 degrees) should have some east-wall glazing to admit the early morning sun, which is particularly beneficial to plants.

2. A sunspace or solarium intended primarily for heat production and additional space need not have glazing in the east and west walls or in the roof. If it is to be built in a region with a cloudy climate, however, it should have some roof glazing to collect diffuse sky radiation from overhead. Unglazed end walls should be insulated to at least R–19.

3. In all climates at least the rear (north) portion of the solarspace should not be glazed and should be heavily insulated (to at least R–30). Because the greatest amount of heat builds up under the highest portion of the roof—the part closest to the house—it will be easily lost to the outdoors if this portion of the roof is glazed and uninsulated. On the other hand, you will want to lose excess heat easily during the cooling season; therefore, you should plan to have exhaust vents, operable skylights, roof windows, or wind turbines in the solid portion of the roof. The solid portion of the solarspace roof also shades the south wall of the house in summer, thus reducing the cooling load of the house.

4. A solarspace attached to a frame wall needs to have additional mass, in the form of masonry, adobe, water or phase change materials in containers, or earth in planting beds or large pots, incorporated into it to prevent extreme fluctuations in temperature within the space (see thermal mass sizing procedures later in the chapter). If the solarspace is to be used primarily for heat production and not for growing food and plants during the winter months, however, the solarspace does not need additional mass.

5. A solarspace that covers the entire south wall, or a distinct section of the south wall of a house, usually appears to be more a part of the house than one that covers only a portion of the wall or section. In general, the larger a solarspace is, the greater the volume of heated air it will supply to the house.

3–1. Two-story solarium covers entire south wall of house designed by Paul Shippee, Boulder, Colorado (Photo: Darryl J. Strickler)

6. Whenever possible, the building materials used in a solarspace should match or harmonize with the materials used for the rest of the house. For example, if the house has wooden siding, the same kind of siding should be used on the exterior of the east and west walls of the solarspace. Similarly, if the house is brick or brick veneer, the same kind of brick should be used on the walls of the solarspace. The roofing materials (shingles or whatever) used on the house should also be used on the roof of the solarspace unless the house has a flat, built-up roof.

7. A one-story solarspace should be 8 to 10 feet (2.4 to 3 m) deep and a minimum of 12 to 16 feet (3.6 to 4.8 m) long. A space smaller than this will not provide adequate room for food production and other living-related functions. A small, low-mass solarspace can, however, work well as a solar heat collector if the heated air is drawn off by fans. A space deeper than ten feet may not allow the sun to reach its back (north) wall.

8. If a sunspace or solarium is to be included in the plans for a new house—or as a part of an overall remodeling plan for an existing house—and if the space is not intended primarily for food production, east, west, and north walls of the solarspace made of solid, uninsulated masonry should be located inside the perimeter of the house.

3-2. Integrated solarspace (Adam residence; Photo, courtesy designer-builder Valerie Walsh, Solar Horizon, Santa Fe, NM)

3-3. A solarspace must have adequate ventilation for summer cooling (Designer: Steve Andrews, Denver, CO)

The heat stored in the masonry walls of the solarspace can then pass through the walls to adjoining living spaces rather than being lost to the outdoors as it would be if the east and west walls of the solarspace were outside the perimeter of the house.

9. All solarspaces, regardless of the type or the climate in which they are built, must have adequately sized openings to ventilate the space during the cooling season. A combination of low and high vents (or exhaust fans) and shading is most effective for cooling a solarspace in summer.

10. The wall between the solarspace and the house must have openings (or fans) located high on the wall so that heated air can flow into the house, and openings located near the bottom of the wall so that cooler air can return to the solarspace by convection. Existing or newly placed windows, doors, or sliding glass doors can serve as high and low vents if they are placed in appropriate locations. All openings between the solarspace and the house must be able to be closed off so that airflow back and forth can be regulated. (Some people make the mistake of removing a section of the wall between the solarspace and the house, thus pro-

viding open access to the space from the house. Do not be one of them.)

11. For optimum heat gain in winter, the south-facing glazing of a solarspace should be set at an angle perpendicular to the winter sun. At 36° north latitude, for example, the sun is approximately 30.5 degrees above the south horizon at noon on December 21. If the south glazing of a solarspace at this latitude were slanted 60 degrees from horizontal, the sun would strike the glazing at a 90-degree angle; thus, it would come straight in. When sunlight strikes a glazing surface that is perpendicular to the angle of

3-4. Sloped glazing on author's integrated solarspace is perpendicular to the winter sun (Designer: Darryl J. Strickler, Passive Solar Design Service; photo: Darryl J. Strickler)

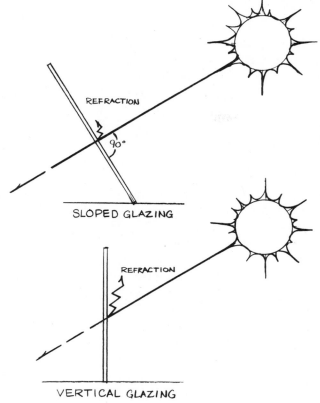

REFRACTION

90°

SLOPED GLAZING

REFRACTION

VERTICAL GLAZING

*3-5. Refraction from sloped and vertical glazing: winter sun angle (36° north latitude)*

the sun, less sunlight is reflected by the surface of the glazing itself. Therefore, the heat gain through the glazing surface is greater than it would be if the glazing were not perpendicular to the sun's altitude. For optimum plant growth, glazing should be tilted at an angle equal to the latitude plus 20 degrees.

12. In cold climates (generally above 40° north latitude), the glazing on the south wall of a solarspace may be vertical without significantly affecting solar gain through the glazing. This is particularly true if the area to the south of the solarspace is frequently snow covered during the heating season, since the highly reflective surface of the snow bounces sunlight into the solarspace. Vertical glazing is, however, not recommended for solarspaces that are to be used primarily for growing space, since plants require maximum light. The framing for a solarspace with vertical glazing is somewhat easier to construct. Moreover, vertical glazing is easier to shade in summer and to insulate at night and during cloudy weather in winter. (In cold climates nighttime insulation of glazing surfaces to prevent excessive

heat loss through the glazing is often cost-effective. It is not always cost-effective in milder climates.)

As suggested in the guidelines above, the actual size and shape of your solarspace should ultimately be related to its intended use, the length of the south wall (or section of the south wall) of your house, and the amount of money available to build it. Beyond these generalizations many specific factors must be considered in planning a solarspace. These include the altitude of the sun during the heating and cooling seasons; the type, amount, and location of thermal mass within the solarspace; the size of glazing and other construction materials selected; and the means of air distribution, ventilation, and shading.

At this point, incorporating all of these factors into the planning of a solarspace may seem like a difficult task. It is certainly a challenging task—the kind of challenge most people enjoy—but it is not difficult. All you need to do is take it one step at a time.

# TEN SPECIFIC STEPS FOR PLANNING A SOLARSPACE

## Step 1: Estimate the Cost Per Square Foot

The first step many people feel they need to take is to find out how much something will cost before they go any further. This is unfortunate because it often places unnecessary limits on creativity and invention. Nonetheless, it is a practical way to begin the planning process—so

*3-6. Vertical glazing of Stockton solarspace, Albany, OR; ample roof glazing not visible in photo (Designer: Charles Bliege, Hyteh, Inc., Sweet Home, OR; photo: Darryl J. Strickler)*

here goes. The cost of your solarspace will depend on what size it is, who builds it, what materials are used to construct it, and how it is attached to, or integrated into, your house. If you construct it yourself on an existing south wall that already has windows and doors and if you use low-cost or salvaged materials, you should be able to build a rudimentary greenhouse for five dollars or less per square foot (0.09 ca). For this price the greenhouse would probably have a gravel floor, 2 × 4 framing, and polyethylene sheeting or fiberglass glazing.

If you opt for an elegant sunspace or solarium designed and built by professionals, it might cost as much as fifty dollars or more per square foot (0.09 ca), depending on how it is built and what materials are used. Within the wide range of five to fifty dollars per square foot (0.09 ca) are many options and trade-offs that you can investigate throughout the remainder of this chapter. Your goal should be to design a solarspace with a per-square-foot cost that is comparable to the value, per square foot, of your house. For example, if you have a 2,000-square foot (180-ca) house worth eighty thousand dollars, the cost per square foot (0.09 ca) of your solarspace should be about forty dollars, or about twenty dollars per square foot (0.09 ca) if you build the whole thing yourself.

It may be helpful at this point to review the examples of existing solarspaces included at the end of chapter 1. Going back to these examples should give you a better idea of what you could accomplish and how much it is likely to cost. Keep in mind that these examples represent region-specific and owner-specific design considerations. Therefore, select examples of solarspaces in regions with a climate and latitude similar to your own region. Also look at examples of solarspaces that are intended for the same purposes you have in mind.

For instance, Ethel Smith's sunspace in Pennsylvania (figs. 1–13 and 1–14) has approximately 240 square feet (21.6 ca) of floor space. Eight double-insulated, tempered glass patio door replacement units were used for glazing, and the structure was built with 2 × 6 lumber on a concrete block foundation wall. The floor is brick with some areas of loose gravel. The total cost of the materials used to construct the sunspace was approximately $2,000.00 in 1981 ($8.33 per square foot [0.09 ca]—for materials only). A sunspace similar to Mrs. Smith's could probably be duplicated for $10.00 to $15.00 per square foot if all of the construction were done by the owner or unpaid volunteers. A similar solarspace constructed by a professional builder would likely cost between $20.00 and $35.00 per square foot.

By assessing your do-it-yourself skills and your financial situation, you can decide who will build your solarspace and how much it is apt to cost per square foot. This, in turn, will help you decide what size it should (could) be.

As suggested earlier, one shortcoming of this cost-per-square-foot approach is that it may prevent you from inventing ways to save money on building costs or from borrowing some money. For example, if you do not have on hand all of the cash you need to build a larger solarspace that would serve a variety of functions, you might consider applying for a home improvement loan (see chapter 4). This may make more sense in the long run, despite interest charges, since a larger solarspace could potentially supply a greater volume of heated air to your house and thereby futher reduce your home-heating costs. In this case the long-term energy savings could offset the interest charges on the loan—especially when you consider that the cost of energy is likely to increase faster than interest rates. Think about it.

## Step 2: Establish the Depth

As stated above, most one-story solarspaces are 8 to 10 feet (2.4 to 3m) deep from north to

3–7. Custom-built solarspace retrofit (Photo, courtesy of designer-builder Valerie Walsh, Solar Horizon, Santa Fe, NM)

south. Although this may seem somewhat narrow, this depth usually allows the low winter sun to illuminate the floor and at least the lower part of the north wall of the solarspace. It is especially important that the sun strike the wall between the solarspace and the house if this wall is constructed of masonry and is intended to store heat.

A solarium with a two-story south glazing area could be as deep as 12 to 15 feet (3.7 to 4.5m). Depending on the specific shape of the solarium, the added height of the south glazing would allow the winter sun to fully penetrate the space.

To establish the optimum depth for your solarspace, first determine how far your property line (or setback requirement, if any) is located from the south wall of your house. Then decide how much of the floor area should be devoted to planting beds, aisles, a dining area, a hot tub, potting bench, storage cabinets, or whatever you have in mind. Add the square footage of the areas required for each function to decide the depth of your solarspace in relation to its length.

## Step 3: Establish the Length

If you have already decided that your solarspace should be attached or integrated along the entire south wall or a predetermined section of the south side of your house, the task of determining the length of the solarspace is easily completed by measuring the wall. If you are planning to enclose a south-facing porch or an existing patio, both the length and depth of the solarspace are predetermined.

A solarspace acts as a buffer zone for the south wall of the house, reducing heat gain and loss through the wall by protecting it from outdoor conditions. Therefore, the more of the south wall you cover with a solarspace, the less the heat loss from the house in winter and heat gain in summer. If the south wall of the house is masonry (brick, stone, concrete block), there is an additional reason to build the solarspace over the entire wall. Heat produced in the solarspace will "migrate" by conduction along that wall and be lost to the outdoors through the unenclosed portion of the wall. Therefore, if you do not cover the entire masonry wall with a solarspace, you should consider insulating the

outside of the unenclosed part of the wall or make a thermal break between the enclosed and unenclosed portion of the wall by removing the masonry to create a gap that can be insulated.

Subtract 1 foot (0.3 m) from the desired length of your solarspace to allow for two 6-inch-thick (15.2-cm) end walls, one on the east and one on the west end of the solarspace. Then divide the remaining length of the wall by 3 feet (0.9 m), and hope that the answer comes out even or very close. The reason for dividing by three is that 3 feet (0.9 m) is an ideal modular size for designing and constructing a solarspace. Patio door replacement glass units are made in a 34- by 76-inch (86.4- by 193-cm) size and are readily available in most areas. Using this size glazing will allow you to use standard "two-inch" lumber—actually 1½ inches (3.8 cm) thick—such as 2 × 6s or 2 × 8s for structural framing of the solarspace. When the framing members (mullions) that hold the glazing are placed 36 inches (91.4 m) apart, center to center, a 34-inch-wide (86.4-cm) glazing unit can be installed *between* them. With the glazing in place, 2 × 6 or 2 × 8 framing lumber will usually provide adequate structural strength for a solarspace.

If you are sizing your solarspace to fit your budget rather than your south wall, you can simply refer back to the information on cost per square foot in step 1, and decide how many 3-foot modular units would yield a space that you could afford.

Patio door replacement glass units are a popular choice for solarspace glazing because they are relatively inexpensive when purchased from a large-volume glass manufacturer. In addition to the 34- by 76-inch (86.4- by 193-cm) size, they also come in 28- by 76-inch (71- by 193-cm), 34- by 90-inch (86.4- by 228.6-cm), 46-by 76-inch (116.8- by 193-cm), and 46- by 90-inch (116.8- by 228.6-cm) sizes. Although any type of double insulated window units can be used for solarspace glazing, patio door glass is usually the least expensive.

If you decide to use patio door glass as glazing for your solarspace and the length of the space you are planning cannot be equally (or almost equally) divided by three, try dividing by 2½ feet (0.8 m) or 4 feet (1.2 m) and plan to use the 28-inch (71-cm) or 46-inch (116.8-cm) size. In this case the mullions of the solarspace would be placed 30 inches (76.2 cm) or 48 inches (121.9 cm) respectively, center to center, if the glazing is installed *between* the mullions.

If you decide to use fiberglass, acrylic, or other synthetic glazing instead of glass, check with the manufacturers listed in the Appendix to determine the sizes of these products so you will be able to design your solarspace accordingly.

## Step 4: Plan the Height, Configuration, and Attachment

After you have decided what depth and length your solarspace will be and what type of glazing to use, the next step is to determine what its height and the angle of its south glazing will be. If you are planning a one-story solarspace, your first decision is whether to set the glazing vertically or at an angle perpendicular to the sun during the heating season. Both approaches have advantages and disadvantages. If the solarspace is over 16 feet deep (4.9 m), you may need to incorporate a load-bearing beam supported by vertical posts where the sloped glazing and the roof of the solarspace meet. A solarspace with vertical glazing is somewhat less difficult to construct, and vertical glazing is easier to shade in summer and insulate at night and on cloudy days in winter. But when vertical glazing is used on the south wall of the solarspace, some roof glazing is usually required, especially in cloudy climates, so the sun can reach the north wall of the solarspace in winter. Roof glazing is, of course, more difficult to install and to keep from leaking, and excessive heat gain through the roof glazing can be a problem during the cooling season if the roof is not adequately shaded. Vertical or sloped? You pays your money; you takes your choice.

The orientation and construction of your house should help you decide on the shape of your greenhouse, sunspace, or solarium and on how it could be attached to your house. (Of

---

**In Praise of Patio Door Replacement Glass: An Editorial**

For openers, I will readily admit that I have a strong bias in favor of using double-insulated, tempered patio door replacement glass units for solarspaces that are not primarily intended for food production. Although many types of fiberglass and acrylic glazing materials are currently available (see Appendix), very few are less expensive per square foot than patio door glass—especially when two layers are used—nor have any been around as long as glass.

A few advantages of patio door replacement glass are:

- it will not discolor or deteriorate from the accumulated effects of ultraviolet radiation and air pollution
- it can withstand many kinds of abuse without breaking, short, perhaps, of a direct blow from a sledgehammer
- it is readily available at an affordable price in most areas of the country

Especially important from my point of view—as I glance up from my writing to look through the glass in my sunspace—is that I can actually see through it. On my house, or any house for that matter, glass looks more natural, in harmony with what the eye *expects* to see when looking out of, or at, a house than do most other types of glazing material. Depending on what outdoor angle you view it from, the glass reflects a mirror image of the sky, a tree, a cloud, or the sun, which is a whole lot more pleasing to the eye than milky or slightly brown-tinged plastic.

The glass replacement units for the triple- and quadruple-glazed line of Solaire Film patio doors now being manufactured by Weather Shield may become more readily available at an affordable price. When they are, buy them; or if you can afford them, buy them now. These specially built units include 3M Sungain solar enhancement film as the middle glazing layer(s) between two sheets of tempered glass in a sealed unit. The Sungain film reduces heat loss through the glazing surface while increasing solar gain. Check the prices of these units when you are ready to build your solarspace. The added cost may be worth the price if the units allow you to get by without insulating the glazing in your solarspace on winter nights.

Having thus extolled the virtues of using patio door replacement glass, it may be well to mention a few possible disadvantages. First, the larger units weigh nearly 100 pounds (45.5 kg) each. Although their weight and size make them

---

somewhat difficult to handle, two or three people working together can usually install them without difficulty. The weight of the units also requires that the framing members and glazing stops of the solarspace be sturdily built to support the weight of the units.

Some designers and builders are hesitant to use patio door replacement glass on an inclined angle. This is because, when the units are set on an angle other than 90 degrees, they tend to sag, or "belly-out," in the center. The 46-inch (116.8-cm) units are especially prone to sagging when they are installed on a slant. Although the tempered glass used in these units usually has sufficient strength to withstand the stress resulting from sagging, wind loads, and differences between the temperature of the inner and outer panes, the seals used to bond the two pieces of glass together around the edges sometimes break. This can result in a buildup of condensation between the two panes, which reduces the transmittance of light (and heat) through the unit. (To balance the picture, I should mention that the same thing can happen to the units when they are installed vertically.) This apparent problem can be avoided by using units that have a special double seal and by taking extra care to build the glazing stops (the surfaces that support the glass)—especially the lower stops—so that they will support the edges of both panes of glass in the unit equally. (The details of how to build glazing stops are contained in chapter 5.)

Despite the potential problems that can result from installing patio door replacement glass units on an angle, I offer the following experience to support the fact that even 46-inch-wide (116.8-cm) units can be used on an inclined angle. My own sunspace, which is nearly 60 feet (18.3 m) long and is glazed with thirteen 46- by 76-inch (116.8- by 193-cm) units set on a 60-degree angle, has experienced wind loads of up to 80 miles (128.7 km) an hour and effective outdoor temperature changes of as much as 100°F (37.8°C) within a single twenty-four-hour period. These units have been in place several years; to date, none of the seals have broken. On the theory that pride precedeth the fall, I humbly realize that by making this statement I could be increasing the likelihood that all thirteen units will crack tomorrow. This is highly improbable, but even if they did, I also realize—perhaps not humbly enough—that I would still be ahead financially since I have already paid for the glass at least three times over through savings on heat bills.

In some instances I have specified fiberglass or double-skinned acrylic glazing for solarspaces I have designed. In one case I had clients who were apparently more concerned with who could see into their sunspace than they were with being able to see out. For their sunspace I used CY/RO Exolite, a double-skinned acrylic glazing material. Although this material is transparent, it has vertical ribs between the two layers that refract light. Looking through Exolite is somewhat like looking through the ribbed glass that is often used as partition between offices. You can see that someone is in there, but you cannot tell who it is or exactly where "it" is.

For another client, I had designed a two-story solarium that was to be set back inside the perimeter of the south wall of the house. The solarium was to have overhead glazing set on a very low angle (30 degrees) to match the pitch of the south roof of the house. Because of the potential problems with setting double-insulated glass at this low angle, we devised a plan to use one layer of fiberglass glazing on the inside of the solarium roof glazing area and a single layer of tempered glass (patio door replacement size, of course) on the outside. The inner layer of fiberglass scattered the light that entered the solarium from overhead. The outer layer of glass gave the solarium a more unified appearance from outdoors since the vertical south wall of the solarium was glazed with the same size of patio door replacement units. (The glazing units in the south wall were, of course, double insulated.) By not using double-insulated units in the roof area, we were able to avoid potential problems with broken seals.

Another major exception to my bias for patio door replacement units is in designing attached solar greenhouses. For clients who are dead serious about food production, I sometimes specify two layers of Tedlar-coated fiberglass glazing (with an air space between them), such as Filon or Sun-Lite. For low-cost greenhouses, I prefer Woven-Poly, a laminated polyethylene sheeting or Du Pont Tedlar polyvinyl fluoride glazing film. The diffusing qualities of these types of translucent glazing spread light and heat more evenly throughout a greenhouse. This helps prevent hot spots, which sometimes result in a bad case of sunburn for plants that are struck directly by sunlight entering a solarspace through transparent glass.

Having noted these few exceptions, I will conclude this editorial as I began it: For my money (or yours), you cannot beat patio door replacement glass units.

course, if you are enclosing a porch, the shape, size, and attachment are predetermined.)

If the roof ridge of your house runs east and west, a portion of the south slope of the roof may overhang the south wall of the house. Depending on the distance of the horizontal projection of the overhang, it may be possible to incorporate it into the roof of the solarspace. If the overhang projects out 3 or 4 feet (0.9 or 1.2 m) from the south wall of the house, it can serve as the only solid portion of the solarspace roof if it can be heavily insulated. Figure 3–8 illustrates how an existing south wall overhang can be incorporated into the roof of a solarspace. Other possibilities for a house that has an east-west roof ridge include extending the overhang to form the insulated portion of the solarspace roof or tucking the roof of the solarspace under the existing roof overhang (see fig. 3–9).

If the roof ridge of your house runs north and south and the south wall is the narrowest side of your house, your solarspace could be attached like one of the examples in figures 1–23, 3–10, and 3–11. A solarspace for a house with a flat roof can simply be attached to the south wall since it does not need to be integrated into, or under, an existing roof slope.

If you are really adventurous, you might consider building a sunspace or solarium *inside* the perimeter of your house—that way you will not need to worry about how to attach it to the house. An integrated sunspace or solarium could

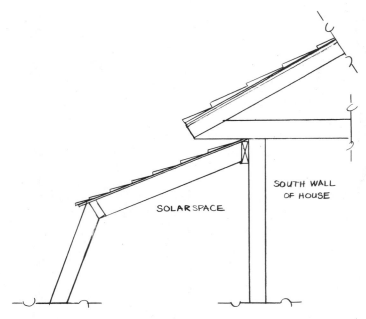

3-9. *Solarspace roof attached to south wall under roof overhang.*

either be built into the south end of a house, as illustrated in figure 3–12, or it could be incorporated into the south wall of an east- or west-facing house. If you consider building an integrated solarspace, follow the same guidelines you would follow for sizing thermal mass, glazing, and ventilation for a attached solarspace.

When you have decided how and where your solarspace will be attached (or integrated),

3-8. *South wall overhang incorporated into roof of solarspace (Builder: Larry Price; photo: Darryl J. Strickler)*

3-10. *Robert and Eileen Ross's solarspace is attached to the south wall of their east-facing house (Designer-builder: Thom Ross; photo: Darryl J. Strickler)*

3-11. The Nussebaum/Wills solarspace faces south; house faces west (Designer: Scott Cummings, SUNERGI, Ashland, OR; photo: Darryl J. Strickler)

3-12. Lines of existing mansard roof extended to enclose solarspace (Photo: Darryl J. Strickler)

measure the exact height it will be at the point where it attaches to the house. Then make your final decisions on the size, type, and angle of the solarspace glazing and the pitch of the roof of the solarspace. These decisions will enable you to determine the shape and configuration of your solarspace.

One of the design problems you may encounter is sizing the height of the south wall of the solarspace so that you can use standard-size glazing. This problem can usually be solved rather easily by adjusting the height of the foundation wall (stem wall) so that enough space will remain between the sill plate on the stem wall and the roof of the solarspace to accommodate the size of glazing you want to use. A possible exception to this kind of flexibility would be in the case where local building codes require a stem wall of a given height. Patio door replacement glass units come in 76-inch (193-cm) or 90-inch (228.6-cm) lengths; fiberglass and acrylic glazing is available in various lengths and can be cut to fit a given requirement (check sizes with manufacturers, listed in the Appendix).

Another way to adjust the height of the south wall glazing is to build a vertical knee wall, a low, framed wall, above the foundation wall. The higher the stem wall or knee wall is, the greater the amount of shaded floor space—especially during winter when the sun is low in the sky. Therefore, it would be a good idea to incorporate awning, sliding, or basement-type utility windows (with double-insulated glass) into the low south wall of the solarspace to reduce floor shading. These windows provide an additional benefit because they serve as intake vents for summer cooling.

A stem wall or knee wall 2 feet (0.6 m) or more in height can minimize the problem of reduced head and shoulder room in a solarspace that has sloped south glazing. This would hold as true for a knee wall built aboveground as it would for a stem wall built below ground level (where the floor of the solarspace is dug into the ground). If you opt for a sunken solarspace, however, it is still a good idea to leave enough of the stem wall aboveground to allow space for intake vents or windows in the stem wall.

## Step 5: Determine How the Sun Will Strike the Space

The next step in your initial planning efforts is to produce a scaled section drawing of your proposed solarspace that shows its depth, height, and shape. This drawing will help you determine how and where the sun will strike the interior of your solarspace and where to place the thermal storage mass. This may sound a bit complicated, but it is not. All you need is a piece of graph paper, a pencil (with eraser), a protractor, a ruler, and your own ideas. Start by drawing a vertical line to represent the height of the south wall of the house. Next draw a horizontal line to represent the depth of the floor of the solarspace; then complete the section view by drawing the configuration you have decided upon (see examples in fig. 3–13).

When you have completed the section drawing, you are ready to add the lines that represent the sun angles. Use the chart in figure

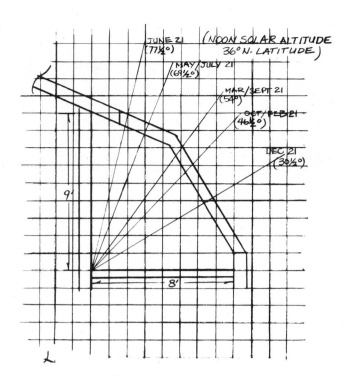

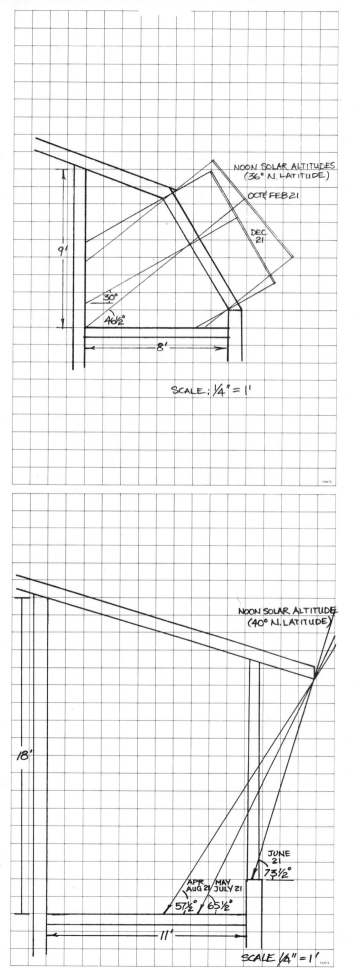

3-13. *Section view of proposed solarspaces*

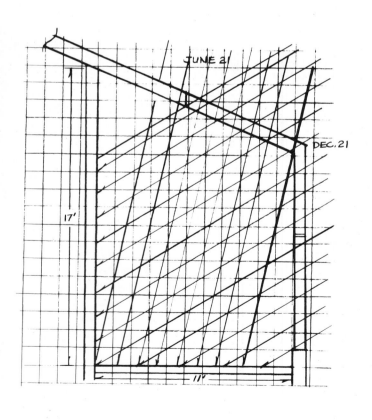

3-14 to determine the altitude of the sun at noon on December 21 at your latitude. Add 8 degrees to the December 21 solar altitude for each month before and after December 21 that your house will need heat. For example, if the peak months of the heating season in your area are from late October through late February, you would add 16 degrees to the sun's altitude on December 21—8° to account for January and November, when the sun is at the same altitude, and 8° to account for February and October again, when the sun is at the same altitude. At 36° north latitude, the altitude of the sun on October 21 and February 21 would be about 46.5 degrees (30.5 + 16). By superimposing lines that are 46.5 degrees and 30.5 degrees (from horizontal) over the scale section drawings, you can estimate how the sun will strike your solarspace at noon during the heating season. You can also estimate how it will strike your solarspace during the cooling season by *subtracting* 8 degrees from the sun's noon altitude on June 21 for each month before and after June 21 that your house does *not* require heat. This will allow you to strategically place thermal storage mass within the solarspace so that it will be struck by the sun during the heating season but not during the cooling season. Keep in mind, however, that the altitudes on the chart in figure 3-14 represent the sun's angle at *noon* on the dates specified. Since the sun is lower in the sky before and after noon, it will penetrate your solarspace differently throughout the day and throughout the year. It is also well to keep in mind that the sun rises south of east and sets south of west on December 21; it rises due east and sets due west on March 22 and September 22 and rises north of east and sets north of west on June 21. Solid end walls in any of these directions will therefore block the sun at the respective times of the year.

## Step 6: Plan the Type, Size, and Location of Thermal Storage Mass

The interior of a well-designed solarspace must have sufficient thermal mass to store heat and prevent large daily temperature fluctuations within the space. If you are designing a solarspace only to produce heat, however, and you do not also want to grow plants or food in the space or use it much in winter, do not be too

3-14. *Altitude of sun (in degrees) above the horizon at noon*

## Altitude of Sun (in Degrees) above the Horizon at Noon

| Latitude | Dec. 21 | Mar. & Sept. 22 | Jun. 21 |
| --- | --- | --- | --- |
| 28°N<br>Orlando, FL (28.5)<br>Tampa, FL (28.0) | 38.5 | 62 | 85.5 |
| 30°N<br>Jacksonville, FL (30.5)<br>Houston, TX (30.0) | 36.5 | 60 | 83.5 |
| 32°N<br>Savannah, GA (32.1)<br>Shreveport, LA (32.5)<br>Tucson, AZ (32.1) | 34.5 | 58 | 81.5 |
| 34°N<br>Columbia, SC (33.9)<br>Wichita Falls, TX (34.0)<br>Los Angeles, CA (33.9) | 32.5 | 56 | 79.5 |

*continued . . .*

| Latitude | Dec. 21 | Mar. & Sept. 22 | Jun. 21 |
|---|---|---|---|
| 36°N<br>Greensboro, NC (36.1)<br>Nashville, TN (36.1)<br>Tulsa, OK (36.2)<br>Las Vegas, NV (36.1) | 30.5 | 54 | 77.5 |
| 38°N<br>Richmond, VA (37.5)<br>Lexington, KY (38.0)<br>St. Louis, MO (38.7)<br>Sacramento, CA (38.5) | 28.5 | 52 | 75.5 |
| 40°N<br>Philadelphia, PA (39.9)<br>Indianapolis, IN (39.7)<br>Denver, CO (39.7)<br>Red Bluff, CA (40.1) | 26.5 | 50 | 73.5 |
| 42°N<br>Boston, MA (42.4)<br>Detroit, MI (42.4)<br>Chicago, IL (41.8)<br>Medford, OR (42.4) | 24.5 | 48 | 71.5 |
| 44°N<br>Portland, ME (43.6)<br>Rochester, MN (43.9)<br>Rapid City, SD (44.0)<br>Salem, OR (44.9) | 22.5 | 46 | 69.5 |
| 46°N<br>Caribou, ME (46.9)<br>Bismark, ND (46.8)<br>Yakima, WA (46.6) | 20.5 | 44 | 67.5 |
| 48°N<br>International Falls, MN (48.6)<br>Cut Bank, MT (48.6)<br>Seattle, WA (47.4) | 18.5 | 42 | 65.5 |

concerned about the amount of mass your solarspace has. Actually, a low-mass space will provide higher-temperature air that can be drawn off to heat the house during a sunny winter day. The interior temperature of a low-mass space will, however, also fall close to the outdoor temperature at night and during sunless days. (For this reason, the wall of the house on which a low-mass solarspace is attached must be heavily insulated. Windows and sliding glass doors between the house and a low-mass solarspace should have movable insulation.) Heat produced by a low-mass space can also be effectively used to charge a fan-forced thermal storage bed under the house (see Remote Thermal Storage Beds, page 52).

If you want to keep interior temperature fluctuations in your solarspace within the comfort range for people and plants—generally, 50° to 85°F (10° to 29°C)—appropriate amounts of thermal mass must be located in the space where it will be in direct sunlight from 9:00 A.M. to 3:00 P.M. during the heating season. Mass materials in a solarspace that are not struck by direct sunlight also help to reduce temperature fluctuations by absorbing excess heat from the air.

Many building materials and strategies can be used to incorporate thermal mass into a solarspace. These include brick, stone, quarry tile, filled or solid concrete block, adobe, and poured concrete floors. Water or phase change material (PCM) in containers, even a hot tub, spa, uninsulated water-heater tank, or fish tanks, can serve as thermal mass, as can large, earth-filled pots.

Some general guidelines may be useful at this point to help you decide what type and amount of thermal mass you should have in your solarspace. The guidelines on the chart in figure 3–15 are based on the ratio of thermal mass to the area of south glazing, so you will first have to calculate the proposed south glazing area of your solarspace in order to use the chart. You will also have to know the average coldest temperature of your area since this also affects the glass-to-mass ratio.

With the above introduction to thermal storage mass, you should be ready to move to such practical matters as how and where to include thermal mass in your solarspace.

### Masonry and Adobe

The combined experience of many designers and owners of solarspaces has indicated that the best all-around solution for a sunspace or solarium is to include as much of the required mass as possible in the structural materials used to build the space. This can be accomplished by constructing an uninsulated north wall of solid (or filled) masonry and the east and west end walls and the stem wall of solid or filled concrete block or adobe with insulation on the *outside*. In addition, the floor of the solarspace could be

3–15. *Guidelines for sizing thermal mass*

| Average Coldest Temperature | | Mass per Square Foot of South-facing Glazing* | | | | | | | |
|---|---|---|---|---|---|---|---|---|---|
| °F | °C | Water | | PCM | | Masonry—solid or filled | | | |
| | | | | | | 8″ | | 4″ | |
| 20 | − 6.6 | .66 | 2.5 | .16 | .60 | .75 | .23 | 1.5 | .46 |
| 10 | − 12 | 1.5 | 5.7 | .37 | 1.4 | 1.0 | .30 | 2.0 | .61 |
| 0 | − 18 | 2.0 | 7.6 | .50 | 1.9 | 1.5 | .46 | 3.0 | .91 |
| − 10 | − 23 | 3.0 | 11.4 | .75 | 2.9 | 2.0 | .61 | 4.0 | 1.2 |

*The figures in this chart are based on the following assumptions:

- The thermal mass is evenly distributed throughout the solarspace
- The thermal mass is in direct sunlight from 9:00 A.M. to 3:00 P.M.

- South-facing glazing includes south wall *and* roof glazing of the solarspace
Italic indicates metric conversion measurement (liters or square meters).

3-16. Serpentine brick retaining wall and planting beds in Andy Zaugg's greenhouse add thermal mass (Concept by President Thomas Jefferson; mason: Steve Kornher; photo: Darryl J. Strickler)

3-17. Masonry interior of Van Winkle greenhouse with added water storage containers (Photo: Darryl J. Strickler)

poured concrete with some areas left open for raised planting beds contained behind masonry retaining walls.

A solarspace that has an adequate amount of masonry evenly distributed throughout the interior is not likely to overheat on sunny winter days. It should also be capable of storing sufficient heat to keep the interior relatively warm at night and during cloudy stretches in winter. During the cooling season, the mass inside a solarspace can also absorb excess heat and moisture during the daytime and can be cooled off at night with cooler night air if the space is adequately ventilated.

Although a solarspace with a massive interior as described above would be ideal, it is not always practical or economically feasible to build this much mass into a solarspace. If the wall on which you are going to attach your solarspace is not already constructed from solid masonry or adobe, consider other options for adding thermal mass to the interior of the space.

### Water

The first option that occurs to many people when they think about thermal storage is one that was frequently used by solar pioneers in the early 1970s: dark-colored, 55-gallon (208-l) drums filled with water. Water-filled drums are certainly an effective, low-cost solution for adding thermal mass, and you may want to use them if you are planning a greenhouse intended primarily for food production. They take up a lot of floor space, however, and tend to make a solarspace look like a World War II oil-storage depot. This is especially true if they are left standing in the open and are not incorporated into a low wall or planting shelf.

Water-filled drums or tall water-filled fiberglass columns are not as effective for storing and reradiating heat as many people believe. Because of their height and large volume of water, warmer water stratifies near the top of the containers while water on the bottom remains relatively cool. The net effect of heat radiated from a containerized column of water is, therefore, much less than would be expected if all the water in the column were as warm as that on top. The problem of temperature stratification can be reduced by tightly sealing the barrels or columns and placing them on their sides.

3-18. Water storage container under planting table in Seawell solarspace (Photo: Darryl J. Strickler)

3-19. Modularized water storage container between studs; glazing shown cut-away (Photo: Darryl J. Strickler)

The above discussion is not intended to discourage the use of barrels or columns of water. In fact, they are very effective for reducing temperature fluctuations in a solarspace. But they simply do not work very well for reradiating stored heat energy back to the space or to plants because the heat is concentrated in one location—the top of the containers. Distributing thermal mass over a larger area of the solarspace interior ensures more even radiation of stored heat throughout the space.

Because of its density, water can be very effective as a thermal storage medium if it is in containers with a large surface-to-volume ratio, such as in modularized containers designed to fit between the studs in a wall cavity (see fig. 3–19). Several manufacturers listed in the Appendix produce this type of water container. Their primary value in a solarspace would be to add mass to a frame wall between the house and the solarspace when built into at least the portion of the wall that is struck by the winter sun. Installing such containers in an existing wall requires removing the inner and/or outer covering of the wall and possibly reframing part of the wall to accommodate a given-size container. Obvi-

ously, including water containers in a new wall that was specifically designed to accommodate them would be much easier.

### Phase Change Materials

In recent years containerized phase change materials (PCMs) have become commercially available for use as thermal storage in solar applications. PCM is a eutectic salt solution, such as calcium chloride hexahydrate, that changes from a semisolid ("frozen") state to a liquid state at a given temperature, usually between 70° and 85°F (21° and 29°C). In the process of melting and refreezing, PCM stores and releases a great deal of heat energy and is about four times as effective a thermal storage medium as the same amount of water.

Containers of PCM come in various shapes and sizes, including rods that are 3 to 4 inches (7.6 to 10.2 cm) in diameter and 2 to 6 feet (0.6 to 1.8 m) in length; "pods," cans, "packs," and floor tiles (see Appendix for list of manufacturers). As is the case with water or masonry, PCM is more effective as thermal mass in a solarspace if it is distributed over a large area rather than concentrated in a single location. For example, a rack of PCM rods or a row of pods could be placed side by side along the north wall of a solarspace where they would receive direct winter sun, or the concrete slab floor could be covered with tiles that contain PCM. Suncraft Limited manufactures Clear Heat tubes that are designed to be suspended from the ceiling of a solarspace, where the greatest heat buildup occurs.

Although containerized PCM is very effective and takes up little space, it is relatively expensive. Furthermore, not all manufacturers of PCM have been able to overcome problems with coagulation of the salt solution and leaking containers over a long period of time. (Clear Heat tubes contain an organic polymer solution that is more stable and less corrosive than eutectic salts.) If you decide to use PCM as thermal storage in your solarspace, check with manufacturers about guarantees and recommended applications for solarspaces like the one you are designing.

### Remote Thermal Storage Beds
The approaches and materials just described add thermal mass to the interior of a solarspace.

Another approach is to include mass in a remote location, such as in a fan-charged rock storage bed. Although adding a rock bed to an existing house is difficult, you may want to consider it if you are planning a new house or an extensive remodel of your present home. For example, the crawl space under an existing house could be used as a rock storage bed if the part of the house above the crawl space has a south wall on which a solarspace could be attached.

Fan-charged rock storage beds have sometimes been placed under the floor of a solarspace. Although this is relatively easy to do in the process of building a solarspace, it should be done only if your primary objective is to keep the solarspace itself warm at night. If a rock storage bed is intended to provide heat to the house, locate it under the part of the house you want to heat. It works like this: solar-heated air drawn off of the ceiling level of the solarspace by fans is ducted to the north end of the rock storage bed and forced across (or through) the rocks, which absorb heat from the air. The air passes through the rock bed and is then returned to the solarspace through vents in the floor of the solarspace, thus completing the loop. Because much

*3–20. Fan-charged thermal storage bed under house*

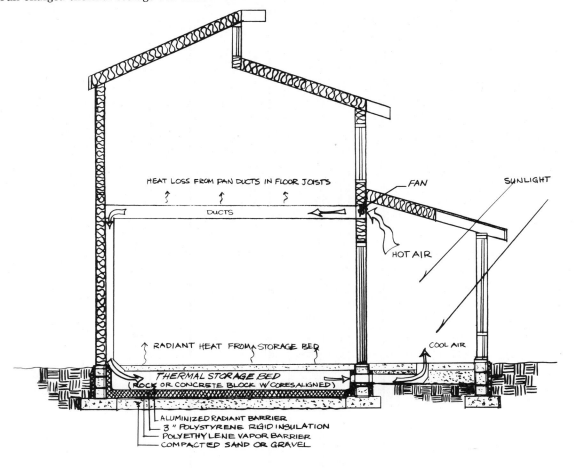

HEAT LOSS FROM PAN DUCTS IN FLOOR JOISTS

DUCTS

FAN

SUNLIGHT

HOT AIR

RADIANT HEAT FROM STORAGE BED

COOL AIR

THERMAL STORAGE BED
(ROCK OR CONCRETE BLOCK W/CORES ALIGNED)

ALUMINIZED RADIANT BARRIER
3" POLYSTYRENE RIGID INSULATION
POLYETHYLENE VAPOR BARRIER
COMPACTED SAND OR GRAVEL

heat can be lost in the trip from the top of the solarspace to the rock bed, ducts carrying hot air to the rock bed should run through living spaces where the heat they give off will not be wasted; or, if the ducts to the rock bed run through unheated spaces, they must be heavily insulated to prevent heat loss. The floor of the house above the rock bed must be uninsulated or have floor registers to allow the heat from the rock bed to enter the living space, and the entire perimeter of the rock bed must be insulated and waterproofed. A rock storage bed should contain 1 cubic foot (0.03 s) of washed, rounded 1½- to 2-inch (3.8- to 5-cm) "river rock" for each 2 square feet (0.2 ca) of south-facing glazing in the solarspace that will supply heated air to the rock bed. (Hollow-core concrete blocks laid on their sides in straight rows, so that their cores are aligned to form air passages, can also be used as remote thermal storage mass in fan-charged systems.) A low-mass or medium-mass solarspace will usually produce hotter air temperatures. Since less interior mass is available to absorb heat in a low-mass space, more of the heat is available to charge a remote storage bed.

## Step 7: Plan Air-distribution Scheme

Openings placed high on the wall between the house and the solarspace allow solar-heated air to move into the house, whereas openings placed low on the wall allow cooler air from the house to return to the solarspace. Although this combination of high and low vent openings will enable the air to flow freely by natural convection, the airflow can be fan assisted if desired. (In fact, a low-mass solarspace should have fans to extract and distribute from the ceiling level the excessive heat that is characteristic of solar-heated spaces that do not have sufficient thermal storage mass.)

When planning for air distribution, first consider using existing windows and doors in the south wall of the house as high and low vents. Decide whether these openings will allow solar-heated air to flow naturally (and quietly) through the areas of the house where heat is needed. For example, if the top of a double-hung window is located high on the wall and can be opened, it can serve as a high vent provided it ad-

mits air to a room that you want to heat. Unless the bottom of the same window is very close to the floor and can also be opened, a low vent would have to be cut through the wall as an air return. (Low vents can be a simple boxed-in opening covered with a standard heating/air conditioning grill or screen and a back-draft damper. Chapter 5 describes how to construct these vents.) When opened, a conventional door or a sliding glass door can serve as both a high and low vent, since air can flow through the top of the opening and return through the bottom.

If your house does not already have enough doors and windows on the south wall, you should consider adding some, since you will also want to achieve a visual connection between the house and the solarspace, as well as a means to enter the solarspace. Sliding glass doors are particularly appropriate because they serve as connections, as well as air-distribution openings.

While you are deciding where to locate windows, doors, or other vent openings between the house and the solarspace, keep in mind that the distance between high and low vents will affect the rate of airflow. Airflow increases with the vertical distance between high and low vents: the greater the distance, the more rapid the airflow. Therefore, vents should be placed as high and low on the wall as possible, but at least 6 feet (1.8 m) apart. If you plan to cut new vents in the wall (or install new windows) try to offset the vent openings so that high vents are not directly above the low vents. This strategy will force heated air to travel a greater distance through the house before it cools and returns to the solarspace.

If you are planning a two-story solarium, remember that heated air must enter the house through openings between the second floor of the house and the top of the solarium. Air passages must also connect the first and second floor of the house—preferably on the north side of the house—so that cooler air can "drop" to the first floor before it returns to the lower level of the solarium. In other words, free airflow between floors—possibly through a stairwell or registers along the north wall—is necessary.

For optimum natural airflow between a house and solarspace, the total area of high vent openings should be approximately 3 percent of the wall area of the wall between the solarspace and the house. For example, if the wall were 8 feet (2.4 m) high and 25 feet (7.5 m) long, the total area would be 200 square feet (18 ca). Thus, about 6 square feet (0.54 ca) of high vent openings would be needed. The total area of low vent openings should also be approximately 3 percent of the total wall area. Thus, the solarspace in the example above would need approximately 6 square feet (0.54 ca) of low vent openings.

To calculate vent sizes, begin by measuring existing door and window openings, then decide if you need to install additional vents. When calculating the vent size of a door opening, count only the top and bottom 12 inches (30.4 cm), since air will circulate only through the top and bottom, not through the entire opening.

If you decide to use thermostatically controlled fans to extract solar-heated air from your solarspace to heat your house—either because it is a low-mass space or because you normally will not be at home to open and close doors, windows, or other vents at the appropriate times—select a fan (or fans) that is capable of drawing about 7 cubic feet (196 l) of air per minute (7 CFM) for each square foot (0.09 ca) of south-facing glazing. (This general rule applies only to situations where there is no natural airflow into the house, such as when no high vents are open.) For example, if you planned to use 200 square feet (18 ca) of south-facing glazing on your solarspace, you would need a (quiet) fan, or fans, that can draw a total of 1,400 CFMs (39,200 l). In this case three 500-CFM fans evenly spaced along the top of the wall between the solarspace and the house would be more effective for distributing heated air over a larger area of the house than one large 1,400-CFM fan. The exception to this advice would be if you wanted to duct all of the heated air from your solarspace to the supply side of a forced-air furnace or to a rock storage bed beneath the house or solarspace. Here, one 1,400-CFM "squirrel cage" fan or an in-line duct fan would be preferable. Remember, if you use fans to extract the heated air, you also need low vents to return cooler air to the solarspace.

You can devise as exotic or as simple an air-distribution system as you wish. For example, you could plan to use vents that are adequately sized to permit natural airflow *and* also install a thermostatically controlled fan system that would be overridden by a trip switch when the manual vents (doors and windows) are open.

With such a setup, your fans will "know" if you are at home and can "decide" whether they will run. You could also use three-speed fans that automatically increase their speed as the temperature of the air in the solarspace increases. Thus, they would run slower and quieter at lower temperatures. Also, you could use an axial fan that can be reversed to blow warm air from the house into the solarspace on cold nights. The possibilities are endless, limited only by your own creativity and your willingness to pay the purchase price and the operating and maintenance costs of mechanical and electronic equipment. But before you get carried away, bear in mind that you can trust air to move by itself if vents are located and sized properly—and that fans make noise and cost money to operate.

If you build a solarspace with adequate thermal storage mass but no one will be at home regularly or no one *wants* to open and close vents, you can install automatic vent operators in high vents instead of using fans. These passive devices, which work without electricity, are also very useful in situations where the high vents to the house are too high to reach conveniently. Vent operators like the Dalen Series 35 Thermal Operator will automatically open and close vents, such as awning windows or a flap vent, at predetermined temperatures. (Automatic vent operators on vents that lead to the house should be disconnected during the cooling season so the warmer air in the solarspace will not enter the house. These same vent operators can then be installed on exterior vents during the summer and reinstalled on the house vents during the heating season.)

## Step 8: Plan Summer Cooling Strategies

In addition to planning for air distribution, you must have a system to reduce the temperature in the solarspace during the months your house does not need heat. The two major strategies used to cool a solarspace are ventilation and shading.

*Ventilation* is accomplished through a combination of high and low vents between the solarspace and the outdoors. Low intake vents should be located in the side of the solarspace that faces the prevailing summer breezes, and

3-21. Wood-frame basement-type windows placed in the stem wall of a solarspace make ideal low vents (Photo: The Solar Project, Lancaster, PA)

3-22. Wind turbines in roof of Nelson solarspace exhaust hot air in summer and are closed off and insulated in winter (Designer: David Marianthal; photo: Darryl J. Strickler)

high exhaust vents should be located on the opposite side or in the roof area. The rate of natural airflow through the solarspace increases as the vertical distance between high and low vents increases. Therefore, air entering the solarspace during the cooling season should be admitted as low as possible—such as through vents in the stem wall, below the glazing—and should be allowed to escape to the outdoors through vents placed high on the walls or ceiling in the solarspace.

Wood-framed awning windows or basement-type utility windows that open fully work particularly well as low vents and can also be used as high vents on the east or west wall of a solarspace. An exterior doorway fitted with a screen door can also be used as an intake vent to admit outdoor air to the solarspace. Operable skylights, roof windows, roof vents, or wind turbines placed in the solid (insulated) portion of the solarspace roof work well as high exhaust vents. The openings for vents in the roof and walls of a solarspace must, however, be heavily insulated and tightly sealed during the heating season to prevent hot air from leaking through them. Wind turbines are especially effective as exhaust vents because they create an updraft when the turbines spin and they can be placed on a stack to increase the vertical distance from low vents. In addition, wind turbines can be easily blocked off during the heating season and fitted with thermostatic dampers.

As an alternative to exhaust vents such as those described above that handle airflow naturally, thermostatically controlled fans can be used to exhaust hot air from a solarspace during the cooling season. (The thermostatic control would, of course, be disconnected during the heating season.) A typical wall-mounted kitchen vent fan or a larger louvered vent fan can be placed high on the east or west wall (whichever is "downwind" from the intake vents), or power roof vents can be placed at the highest point of the solid portion of the solarspace roof. Wind turbines can also be fitted with electrically operated duct fans that are installed inside the stack that holds the turbine. Select a two- or three-speed exhaust fan that can draw 5 to 7 CFM (140 to 196 l) per square foot (0.09 ca) of south-facing glazing (roof and wall glazing).

3-23. Thermostatically controlled fan exhausts hot air from the Seawell solarspace in summer (Photo: Darryl J. Strickler)

For natural ventilation the combined area of intake and exhaust vents between the solarspace and the outdoors should total approximately 15 percent of the floor area of the solarspace. The total area (or volume of airflow) of exhaust vents should be larger than the total area of intake vents. A ratio of two to one works well in most cases.

A good general ventilation plan would be to include appropriately sized intake and exhaust vents to handle natural airflow for summer cooling but to make provisions for electric exhaust fans if you later discover that you need them. For example, as you are constructing your solarspace, frame in a section at the top of the east or west wall that is the correct dimension to accommodate a particular exhaust fan, and run wiring inside the wall that can later be connected to the house current. Instead of purchasing, installing, and hooking up the fan, build an insulated flap vent for the boxed-in section. Similarly, you could frame in and run electricity to areas of the solarspace roof where you may want to add power roof vents later. An exception to this wait-and-see plan would be when you are going to install a spa or hot tub in your solarspace: you will almost certainly need an exhaust fan to remove excess heat and moisture in summer. An air-to-air heat exchanger operated by a humidistat is also worth considering for use in winter, especially since it can remove humidity without throwing away heat. A small bathroom model heat exchanger that sells for less than $150 is adequate for most solarspaces.

A matter of primary importance in determining the location and size of cooling vents is the direction of prevailing summer breezes that strike your house. If your solarspace covers the entire south wall of your house and the prevailing breeze is from a southerly direction, the intake vents in the solarspace must admit an adequate volume of air to cool the house as well as the solarspace. They must, therefore, be larger than would be necessary to cool only the solarspace. The direction of the nighttime breezes is particularly relevant, since you may want to open your house at night and in the early morning during the summer and close it during the hotter parts of the day. Being able to close off the solarspace from the house allows you to keep the house closed during the daytime while the cooling vents in the solarspace are open.

*Shading* a solarspace from the sun to reduce solar gain during the cooling season is obviously very important. Tall deciduous trees to the south, east, and west of a solarspace work very well for natural shading since their leaves appear at just about the time of year that you no longer need heat from the solarspace. Deciduous trees also lose their leaves at just about the time you begin to need heat again in the fall. Pretty neat—unless your trees, like some oak trees, hold their dead leaves over the winter. To effectively shade a solarspace, a tree's canopy must block the sun from striking the glazing surface of the solarspace throughout the cooling season but not during the heating season. Keep in mind that the trunk and leafless branches of trees to the south will also provide some shading even during winter.

Deciduous vines can also be used to provide natural shade. They can be trained to grow on a trellis above the vertical south glazing of a solarspace or to climb a vine arbor placed in front of the glazing. If the south glazing of the solarspace is sloped, plants such as runner beans can simply be allowed to grow up the glazing

3-24. Deciduous trees provide adequate shade to the interior of the Harrison solarspace in summer; note PCM thermal storage rods used as edge on planting beds (Photo: Darryl J. Strickler)

3-25. Bean vines shade the glazing of the Dunnell solarspace (Photo: Darryl J. Strickler)

3-26. Fixed louvered overhang and wires for deciduous vines on Metz solarspace (Designer: Gene Metz; photo: Darryl J. Strickler)

surface or on the mullions between glazing units during the cooling season. Such plants will wither and die when cold weather returns. (Check with a local nursery or commercial greenhouse to find out what kind of vining plants will grow best in your area.)

You might also think about planting some foliage or flowers near the low intake vents of your solarspace after it is completed. If the ground in front of the vents is shaded, the air admitted through the vents will be slightly cooler than it would be if the ground were bare. Planting fragrant plants such as lilies of the valley or gardenias will add a pleasant scent to the incoming air. (Lilies of the valley do not like direct sunlight, so you may need some taller plants to shade them.)

If you are not fortunate enough to have tall trees that will shade your solarspace in summer and you are not wild about the idea of having vines dripping all over your solarspace, you have other options. If the glazing is vertical, you can plan to use a fixed roof overhang above the glass, which will provide shade during the cooling season. The overhang can be solid, or it can be made with fixed or movable louvers that run horizontally and block the sun from striking the glazing in summer but not in winter (see fig. 3–26). If you decide to use a fixed overhang, use the chart in figure 3–27 to calculate the horizontal distance the overhang should project beyond the glazing.

Interior shading devices, such as conventional drapes, insulated curtains, or roll-up or venetian blinds, will certainly block light from striking the walls and floors of a solarspace, but

with these devices, light and heat have already penetrated the glazing before they are intercepted. A considerable amount of heat can build up between the interior of the glazing surface and a shading device placed inside the solarspace. Therefore, you should plan to shade the exterior surface of the glazing.

For exterior shading on sloped or vertical glazing, the best shading strategy other than natural shade is to use a roll-down bamboo or match-stick curtain or a product known as greenhouse shade cloth on the *outside* of the glazing. Shade cloth is a woven fiberglass, polypropylene, or other synthetic material that resembles insect screens. It comes in a variety of colors, with various light-blocking capabilities (see Appendix for manufacturers' addresses). Shade cloth that blocks 70 to 80 percent of the incoming light usually works best because some light is necessary for plant growth and people's vision. If you have any serious crops growing in your solarspace during the cooling season, however, select a shade cloth that blocks no more than 25 percent of the light. Although bamboo or match-stick curtains can be used as external shading devices, they tend to look pretty ratty after being out in the weather a season or two, whereas shade cloth has a much longer life expectancy (For example, VIMCO Solar Shields are guaranteed against deterioration for ten years.) For the money, shade cloth is by far the best investment for shading a solarspace.

3-27. *Shade determination chart*

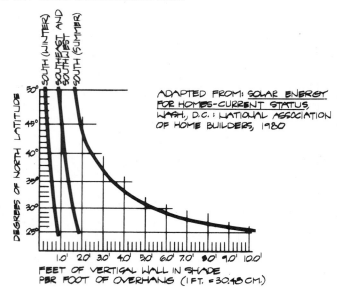

ADAPTED FROM: SOLAR ENERGY FOR HOMES-CURRENT STATUS, WASH., D.C.: NATIONAL ASSOCIATION OF HOME BUILDERS, 1980

FEET OF VERTICAL WALL IN SHADE PER FOOT OF OVERHANG (1 FT. = 30.48 CM.)

USED WITH PERMISSION OF THE NATIONAL ASSOCIATION OF HOME BUILDERS, 15th & M STS., N.W.; WASH., D.C. 20005

## Step 9: Plan Wiring, Plumbing, and Lighting

If you think you will want to have a domestic hot water preheat system (see step 10), hot tub, spa, utility sink, electrical outlets, lights, telephone, or other fixtures in your solarspace, this would be a good time to determine how these items can be tied into the existing plumbing and wiring in your house. The specific fixtures, appurtenances, or amenities you have in your solarspace will, of course, depend on how you plan to use the space and what you can afford.

You should have at least one overhead light for every 150 square feet (13.5 ca) of floor space; a light outside the outdoor entrance to the solarspace; an electrical outlet; and if you are going to grow food, a water faucet and drains in the floor. Track lights installed along the structural framework of the solarspace—such as between the insulated and glazed portion of the roof—will allow you to direct light wherever you need it, but a simple ceramic fixture or warehouse light with a naked bulb will do the job about as

3-28. Greenhouse shade cloth on exterior of solarspace on earth sheltered house (Photo, courtesy of designer-builder: M. S. Milliner Construction Company, Inc., Frederick, MD)

well. Recessed lights in the insulated part of the ceiling are not recommended because they reduce the insulating value to the roof.

Fans located in the walls or ceiling of a solarspace or an air-to-air heat exchanger should be permanently wired into the household current and have a separate circuit breaker or fuse. The wiring must therefore be placed between studs and rafters before the walls and ceilings are insulated. Separate circuits for any pumps or other electrical equipment used to heat or circulate water for a hot tub or spa are also desirable.

Having a hot tub or a spa in a solarspace presents several major contradictions for some people. It not only requires a considerable amount of nonrenewable energy to heat and circulate the water, but the pumps and swirling water constitute a form of noise pollution inside a solarspace. Moreover, you can spend nearly as much for a top-of-the-line spa and related equipment and decking as you would to build the solarspace that encloses it. Despite these contradictions, if you want a place to soak your tired body, you should have one; but before you go the more conventional route, consider some alternatives, such as heating the water with the solar or a wood-fired water heater (see the Appendix for a listing of manufacturers). Or, rather than using a conventional hot tub or spa, install a small

wooden Japanese soaking tub, half of a large whiskey or wine barrel, a small, round concrete septic tank, an industrial-type vat made of hard plastic, an old-time bathtub with feet, or anything else that holds enough water and looks decent. Any of these containers can be built into decking or, better yet, a stone wall and can be made to look very attractive. Furthermore, they can be filled with solar- or wood-heated water when you actually want to use them and left full between uses to serve as added thermal mass and provide moisture to plants. Such a setup would, of course, require a drain, but instead of dumping the water into the sewer or septic system, you could develop an underground watering system for your planting beds (as long as you use only organic or herbal biodegradable cleansing agents in your bathwater). Rather than spend a fortune on a hot tub or spa and its associated equipment, maintenance, and operating costs (unless you happen to have a fortune), why not think about some of the alternatives above and put the money you save to better use—perhaps for something that will keep you warm or produce electricity, rather than waste it.

3-29. Septic tank, hot tub and wood-fired water heater in Goble solarspace (Designer: Akira Kawanabe; photo: Darryl J. Strickler)

### Air-to-Air Heat Exchangers

This would also be a good time to consider adding an air-to-air heat exchanger to your solarspace when it is completed. The energy-saving advantages of sealing a solarspace (or an entire house) very tightly have been well documented. Greatly reduced air-infiltration rates can cut heating and cooling costs dramatically. A tightly sealed space can, however, create problems with indoor air pollution and excess humidity. High-quality air is obviously important to inhabitants of a space, since neither plants nor people thrive well in stale air. Opening an exterior vent to admit fresh air to your solarspace may work very well during warmer weather, but it would not be such a great idea in the dead of winter, since much valuable heat would be lost and a rather nasty draft would be created. A heat exchanger can help ensure a supply of fresh air to the interior of the space while removing pollutants and excess humidity from the air.

As the name implies, a heat exchanger is capable of exchanging indoor air for outdoor air. Most exchangers recover 60 to 80 percent of the heat (or cool) from the air before it is exhausted from the space and transfer that heat (or cool) to incoming air. As an added benefit, an air-to-air heat exchanger also exhausts excess humidity, which is especially important if you will have a spa or hot tub in your solarspace.

Small, window-size units that resemble a window-size air conditioner sell for $150 to $350 and can be installed in a window or through a wall (see Appendix for manufacturers). Although these smaller units are not adequately sized to condition an entire house, most are large enough to provide fresh air and dehumidify an average-sized solarspace. They operate automatically when the humidity in the space reaches a predetermined setpoint. Most models run quietly and efficiently, and because they do not operate continually, they do not use a large amount of electricity on a yearly basis. A heat exchanger is most needed in winter, since exterior vents to the solarspace will usually be open during warmer weather.

By selecting a specific heat exchanger when planning your solarspace, you can make provisions for installing the exchanger and wiring into the household electrical circuit when the solarspace is completed. If you decide on a window unit, be sure to include an appropriately sized window in the solarspace, as well as an electrical outlet nearby. If you opt for installing one through, for example, an end wall, plan to frame in an opening of the correct size and stub in the electrical connection before you enclose and insulate the wall.

If you are not experienced in plumbing and wiring or if your building codes require that such work be done by professionals, have a plumber and electrician help you plan how the utilities will be connected before you start building your solarspace.

## Step 10: Plan an Integral Water Heater

Now that you have planned the basic features and utilities for your solarspace, you may want to consider incorporating some type of solar preheater for your existing domestic hot water system. Very simply, this means that you could include a water tank or collector fins or pipes behind the glazing on the inside of your solarspace. Solar-heated water produced in this manner can either be used directly or it can supply preheated water to a conventional electric or gas-fired water heater. Each of the systems described below can be built for three to five hundred dollars, more or less, depending on the materials used and on who does the installation. The payback period for these systems in terms of energy savings for hot water heating can be as short as three years.

### Collector Fins

Collector fins are 8-foot (2.4-m) lengths of extruded aluminum, 8 inches (20.3 cm) wide, with a length of copper pipe snap-fitted to their black front surface. Three or more fins may be attached to the mullions on the inside of the south wall or roof glazing of a solarspace and connected to a storage tank that is placed higher than the fins. The bottom of the tank should be at least 18 inches (45.7 cm) higher than the top of the fins to prevent reverse flow. As the sun strikes the black surface of the fins, heat is transferred to the water running through the pipes in the fins. The heated water in the fins then expands and rises by natural convection (without mechanical assistance) to the storage

# COLLECTOR FINS

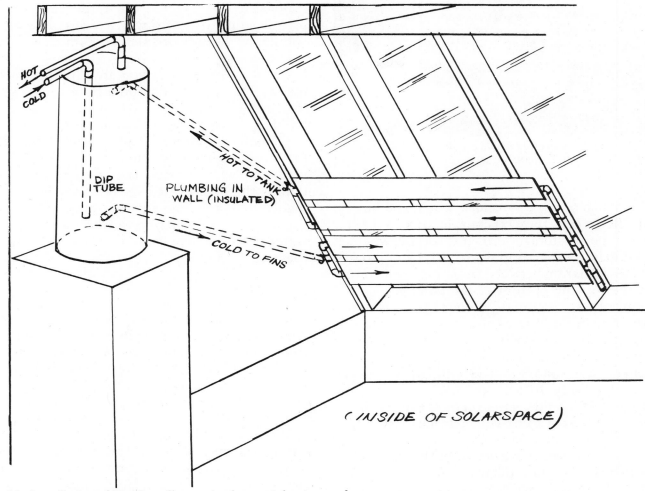

*3-30. Installation of Big Fin collectors in thermosiphoning mode*

tank. The hotter water entering the top of the tank displaces cooler water at the bottom of the tank, thus completing a convection loop. Hot water is drawn off the top of the tank for direct use or to provide preheated water to the conventional domestic hot water heater.

Figure 3-30 illustrates how a simple thermosiphoning collector fin system should be installed. An inexpensive check-valve, to be used when the storage tank cannot be placed higher than the fins, is available from Zomeworks (see Appendix for address).

Collector fins are particularly appropriate for use in a solarspace with a large glazing area. Because they intercept some of the incoming sunlight, they will shade the floor area behind them. Although people in a hot tub located behind collector fins might appreciate the shade and added privacy provided by the fins, vegeta-

tion in a planting bed shaded by the fins probably would not appreciate being in their shadow. Therefore, you should carefully plan fin location to enhance the intended uses of different areas of the solarspace.

Collector fins should be installed where they will receive maximum sunlight year-round. Since hot water is needed all year, be sure that the fins are not shaded by summer foliage or a roof overhang. Fins can either be attached to the back side of the mullions or rafters that hold the glazing or they can be notched into the mullions or rafters as illustrated in figure 3-31. Notching the fins so that they can be rotated will allow you to adjust their angle so that they can be set perpendicular to the sun at different times of the year. It will also enable you to open and close the fins as if they were large venetian blinds, thus providing privacy when you need it.

3-31. *Big Fin collectors inside glazing; note gardenia bush in front of intake vents of solarspace (Photo: Darryl J. Strickler)*

When you have decided where and how to install the fins, the next consideration is the size and type of storage tank to use and where it should be located. A low-profile tank will result in less variation between the water temperature at the top and bottom of the tank. A recycled conventional 30- or 50-gallon (114- or 190-l) hot water heater can be used as a storage tank if it has four fittings—two on top, one on the side near the top, and one near the bottom. As a general rule, one 8-foot (2.4 m) fin should be used for every 10 gallons (3.8 l) of water the storage tank holds. This very general guideline can be adjusted to fit your particular situation without too much loss of efficiency.

The efficiency of a collector fin system will be increased to some degree if the storage tank is located where it will be struck by direct sunlight. In this case an uninsulated tank should be used, and it should be painted black or covered with an adhesive-backed selective surface, such as black chrome. A selective surface like Sunsponge or Maxorb increases heat gain on the outer surface of the tank while reducing heat loss from the tank. If you plan to use a recycled water heater as a storage tank that will be placed in direct sunlight, remove the insulating jacket (usually white outer covering) from the water heater.

An uninsulated tank in a solarspace can serve as thermal storage mass and can help reduce temperature fluctuations in the space. The heat that is lost from the tank is given up to the solarspace, which is an advantage during the heating season but a disadvantage during the summer months.

If you prefer not to have a black tank in your solarspace for aesthetic or other reasons, you might consider placing the tank in a more remote area, such as a closet or attic. If the tank is not struck by direct sunlight, the tank and the pipes leading to and from it that carry hot water should be heavily insulated. Remember that water is very heavy—about 8.5 pounds (3.8 kg) per gallon (3.8 l), so if you are thinking about placing the storage tank on a floor, make sure the floor can carry the extra weight or plan to build additional support for the tank.

Finally, if you are going to use a collector fin system to supply preheated water to your conventional water heater, the storage tank containing the solar-heated water should be placed higher than the conventional tank. This will usually allow the solar-heated water to reach the water heater by gravity flow, thus eliminating the need for a pump.

### Collector Pipes

If your solarspace will have roof glazing and you are seeking a simple, low-cost solution to preheating water, you might consider attaching continuous, interconnected lengths of 1½-inch (3.8-cm) black, high-molecular, polyvinyl chloride (PVC) pipe to the underside of the roof rafters beneath the roof glazing. Without a storage tank, the amount of preheated water would be limited to the volume of water in the pipe at a given time. Since the water in the pipe will, however, heat fairly rapidly on a sunny day, such a system would supply small amounts of hot water throughout the afternoon at least. The addition of a storage tank to the system would, of course, increase the supply of hot water and extend the hours that it is available.

A collector pipe system works on the same principle as a collector fin system, only instead of the fins absorbing sunlight, the pipe itself is the absorber surface. Thus, the storage tank in a collector pipe system should be placed higher than the pipes (in the attic of the house, for example), or a check-valve or pump must be used as described for collector fins.

### Batch-Type Water Heaters

A batch water heater consists of one or more cylindrical water tanks placed horizontally (on their side) or on an incline in a nonfreezing environment where they will be struck by direct sunlight. The tanks are painted flat black or coated with a selective surface to increase absorption so that the sun will heat the water inside the tank.

3-32. Breadbox-type water heater (Photo, courtesy Tennessee Valley Authority, Solar Applications Branch)

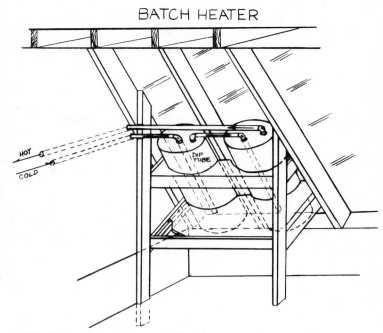

BATCH HEATER

HOT
COLD
DIP TUBE

3-33. Schematic of two-tank batch system

Batch-type heaters have been used all over the world for many decades. "Breadbox" heaters, a type of batch water heater first developed by Zomeworks in 1974, are usually constructed as freestanding units. They typically consist of one or two black tanks enclosed in an insulated box with one or more glazing surfaces that allow sunlight to strike the tank(s), and an insulating/reflective lid, or cover, that is placed over the glazing at night to reduce heat loss. You could use a freestanding breadbox heater or integrate one into your solarspace. (Plans to build a freestanding breadbox heater can be ordered from the sources listed in the Appendix.)

Because a well-designed solarspace (like yours) should not be subject to freezing, it provides the equivalent of the box and the glazing components used in a breadbox water heater. Therefore, all you need to do is place the tank(s) in the interior of the solarspace at a location where they will receive direct sunlight year-round, then connect the tank(s) to the household plumbing system. The hot water from the batch heater can be used directly (to fill your soaking tub, for example), or it can supply preheated water to your existing domestic hot water system.

Two 30-gallon (114-l) tanks can be placed on an inclined angle behind one of the glazing sections in the south wall of a solarspace, or a specially designed support can be built to integrate the tanks under the roof glazing. When two tanks are used, the tanks should be hooked up in a series so that cold water arrives at the bottom of the first tank through a dip tube and is drawn off the top of the second tank (see fig. 3-33). The tanks inside a solarspace may be enclosed in an insulated box with a reflective interior, or they may be left unenclosed. (To build an integrated batch heater, follow the instructions in plans for a breadbox heater, but skip the box part. See Appendix for supplies of plans and materials to build a breadbox heater.)

## GETTING IT ALL TOGETHER

Up to this point, you have been considering each of the major factors that should affect the planning of your solarspace. It is now time to put them all together to design a greenhouse, sunspace, solarium, or enclosed porch. This should not be difficult if you remember that planning a solarspace does not mean that you must be able to produce professional-quality construction drawings. It means that you have made the necessary decisions, worked out an overall plan, and made at least some rough graphic representations or sketches of what your solarspace will look like. If you need blueprints or construction drawings, have a professional do them *after* you have finished all of your planning. For now, keep working and reworking your ideas until everything falls into place. It will, but do not hurry—you will have to live with whatever you plan to build for a long time.

# FOUR
# Finding Help

Who will build your solarspace? Where can you find help to complete the design and construction and pay for it?

If you have the skills to complete the design of your solarspace and can build and pay for it without help, you can skip this chapter. If you do not, read on. This chapter should help you decide what assistance you need and where to find it.

## COMPLETING THE DESIGN

If you have completed all of your planning but you still do not feel you can finish the design of your solarspace, you may want to look for some help. But before you do, make sure you have at least clarified to yourself the functions you want your solarspace to serve. Unless you know how you want to use the space, it will be difficult for anyone else to design it so that you will be satisfied with it.

The search for a designer or architect who is competent and experienced in solar design work may prove to be time-consuming but should be well worth the effort. Start by asking your friends and acquaintances if they can recom-

mend anyone; or, if you are bold, stop and talk to the owners of a solarspace you happen to see on your way home from work. If these endeavors fail to turn up any leads, try the Greenhouse or Solar Energy categories in the Yellow Pages. Chances are that someone in your area, perhaps even a remodeling contractor, specializes in the design or construction of solarspaces or both.

When you have located one or more designers who inspire your confidence, arrange to have them visit your home to discuss what you have in mind. The initial consulation and estimate should be free of charge and should include a discussion of the designer's previous work—including names of previous clients you could call or visit—and the fees charged for the design work. When you have decided what assistance you need and exactly what you want the designer to do for you, arrange to have the design work completed within a specified time period and get an estimate of how much the work will cost. Be sure to inform the designer of any building codes, covenants, or restrictions that apply to your property.

If you deal with a designer/builder, you should also be able to get an estimate for the total cost of materials and labor needed to complete the solarspace if you do not plan to build it yourself. Be aware that some designer/builders have a tendency to build what is easiest to build rather than what will work best. Therefore, it is to your advantage to hire an independent solar designer or architect and then hire another person to build the solarspace.

Should you have difficulty locating an experienced designer or architect, contact your state or local solar energy association. Most of these groups have members knowledgeable in the design and construction of solarspaces who are willing to provide assistance—sometimes free of charge. If no one on the staff of the association does design work, the association may be able to recommend a designer. Or get in touch with the Conservation and Renewable Energy Inquiry and Referral Service (CAREIRS), P.O. Box 8900, Silver Spring, MD 20907, telephone, toll free, (800) 523-2929, from Pennsylvania (800) 462-4983, from Alaska and Hawaii (800) 523-4700; or the National Center for Appropriate Technology (NCAT), P.O. Box 3838, Butte, MT 59701, telephone, (406) 494-4572. Both of these organizations maintain files of resource people in various parts of the country who could help you design and construct your solarspace. They also provide free information and advice.

## CONSTRUCTION DRAWINGS (BLUEPRINTS)

If a building permit is required where you live or if you are planning to seek a loan to construct your solarspace, you will almost certainly need some kind of drawings that represent what you are planning to build. Although formal construction drawings or blueprints may not be required to secure a building permit or loan, you may want to have a solar designer do the basic design work and have a draftsperson do the actual construction drawings. (A good designer may charge forty dollars or more an hour, whereas a draftsperson may charge as little as ten dollars an hour.)

Construction drawings can be useful for getting bids for labor and materials, and they can serve as a sort of official document in case any questions or problems related to the design or construction of your solarspace should arise. If you have construction drawings done, be sure they contain sufficient details to show how the solarspace should actually be constructed. These details will be especially useful to you or a builder who may not be familiar with some of the construction techniques used for solarspaces. (The details in chapter 5 should prove helpful to you or your draftsperson or builder.) Be aware, however, that not every construction detail can be drawn and that some procedures will have to be worked out during the construction process.

## WHO WILL BUILD YOUR SOLARSPACE?

When you have arrived at a final design for your solarspace, the next major considerations are how much of the actual construction you want to do yourself and where to find help if you need it. If you have previous experience with building and do-it-yourself projects, you will probably be inclined to do at least some of the work on your solarspace. If you are inexperienced but willing to learn new skills, you can attend night-school classes at a local vocational-technical school, sign up for a course at an owner-builder school (see Appendix), or participate in a greenhouse construction workshop sponsored by a local organization. A number of excellent how-to books that can help you learn basic construction techniques are also available. (Some of these are listed in the Appendix.) Any or all of these activities will help you gain the skills needed to construct a solarspace and help build your confidence at the same time.

If you are a confirmed klutz with ten thumbs, you have probably long ago abandoned all hope of building anything more complex than a pile of nursery school blocks. Objective self-evaluation is certainly very important, but do not sell yourself short. This could be your big chance to improve your do-it-yourself image and save money by learning new skills. (Perhaps when you receive a few estimates on the cost of having someone else build your solarspace, your motivation to learn these skills will increase.) Most important, the feeling of accomplishment you and members of your household can derive from doing some or all of the construction work is likely to increase your satisfaction and pride in the finished product.

At this point a few cautions on building your own solarspace may be in order. First, be aware that obtaining a home improvement loan could be more difficult if you are doing all of the construction work. Some lending institutions routinely refuse large home improvement loans to do-it-yourselfers; others are more flexible and could be convinced—especially if you can demonstrate your competence to the loan officer. When you are building your own solarspace and paying for it as you go, you may run out of time, money, or both before you complete the project unless you plan very carefully. So before you start building, be sure you have at least enough time and money to enclose and insulate your solarspace so that it will operate effectively and begin to recover your investment through heat and food production. The finish work, such as trimming out the interior or adding a quarry tile floor over the concrete slab, can wait if necessary, but it is important that you finish the solarspace to the point where it will be capable of producing heat and food.

A surprising number of owner-built solarspaces around the country—including my own, I must admit—have been "almost finished" for years. Apparently the motivation to spend additional time and money for trimming out a solarspace decreases greatly for some people after they have completed enough of the solarspace to get it working properly. We do not need any more unfinished solarspaces in this country so plan carefully!

The trade-offs between building your own solarspace or having someone else build it are pretty straightforward. If you have the time, inclination, and talent to build it yourself, you can reduce the total cost of the project by as much as one-half. If you are not so inclined or you do not feel you want to take the time to learn new skills, you obviously have to find someone who is able and willing to build it for you. This usually means you will spend less time but more money. Instead of using money, you may be able to trade services with a friend or someone in a barter organization who has the tools and talent needed to construct your solarspace. Failing this

kind of arrangement, you are left with one basic option: pay professionals to build it.

# WHO WILL SERVE AS CONTRACTOR?

Although the titles Contractor and Builder are sometimes used interchangably, a person who is a licensed contractor may or may not do any of the actual construction work on a specific project. A contractor's primary role is to manage the project from start to finish, make sure materials and workmanship are satisfactory, and have the project completed according to the plans and within the time schedule agreed upon. The contractor is also responsible for ordering materials and hiring subcontractors who will complete various parts of the project. Although a building contractor may employ a foreman to supervise the work, final responsibility for the project rests with the contractor. Of course, all of this comes at a price—usually 15 to 20 percent of the total cost of the project.

## Being Your Own Contractor

Whether or not you plan to do any of the actual construction work yourself, you should seriously consider serving as your own contractor. In addition to cutting the cost of the project, being your own contractor makes a lot of sense for other reasons. Since you will have to live with your completed solarspace, the final (final) responsibility is yours anyway, so why not accept it from the start? You also know what you want to accomplish better than anyone else ever could. (After all, you planned your solarspace, right?)

The following "job description" may help you decide whether you want to serve as contractor for your project. Being the contractor will require time and effort along with some skill in interpersonal relations. You will need to line up subcontractors, such as carpenters, concrete workers, electricians, and plumbers, to do whatever you do not plan to do yourself. This means that you must first identify such people, then talk to them about the work you want them to do, schedule their time, draw up a contract, and order materials. You will also need to be at home long enough each day while the work is in progress to make sure the subcontractors know what they are to do and to inspect their work to be sure it is satisfactory. As the owner-contractor, your role is not to tell workers *how* to do their job, but to be absolutely certain they know *what* you want them to do and how you want it to look when it is finished.

The above description may sound like a full-time job. It is—for at least the period of time it takes to complete your solarspace. You should probably plan to take some time off if your job schedule is not flexible enough to allow you to complete these contractor tasks while you are working full-time.

If you are considering being your own contractor, check your insurance coverage and legal liabilities in case anyone who works for you (including a friend) were to get injured. Subcontractors should have their own insurance coverage for themselves and their employees, or they should sign a waiver drawn up by a lawyer that protects you from claims for medical payments and damages if they are injured on the job through no fault of yours.

## Hiring a Contractor

Most contractors prosper or perish on the basis of their reputation for good work at a fair price. Such a reputation can take years to establish and only a few major mistakes to destroy. If you are not familiar with the work of remodeling contractors in your area, check around—chances are that someone you know knows someone who does good work. Try to find a licensed and bonded remodeling contractor who specializes in solar additions. The experience that such a contractor has gained through previous work of this type will be invaluable to you. Perhaps the contractor has already worked out some of the construction details, such as how to install roof glazing, that can improve your project.

When you have located at least two contractors, invite them to your house and explain what you want to do. Have them come at the same time if you want to save time and repetition, as well as increase their desire to be awarded the contract. Give the contractors a copy of the working drawings or sketches of your solarspace, and ask them to prepare an estimate for the cost of materials and a separate estimate for the cost of labor for each part of the project, including their own fees. With separate estimates for materials and labor, you may be able to decide which jobs are worth doing yourself and how much you will save by doing them. A lump-sum estimate will not allow you to sort out the actual cost of each component of the project. Furthermore, if the contractors doing the estimates have no previous experience building solar additions, they may tend to provide a conservative (high) estimate to protect themselves. A cost-per-square-foot estimate based on the contractor's previous experience in building more typical additions that include carpeting, paneling, plumbing, and Sheetrock, for example, will not apply to a solarspace. This is why the actual cost of materials and labor should be estimated separately.

After you have received estimates from at least two contractors and have inspected some of their previous work (and talked to former clients), you should be able to decide with which contractor you want to work. This decision should be based on more than the dollar amount of the estimates. Select the contractor whose work you like best and in whom you have the most confidence. Also decide on the basis of which person you believe you could work with most effectively and enjoyably. Keep in mind that the contractor you select will affect your life, not only during the construction process, but for as long as you live in your house.

## Hiring Subcontractors (Subs)

If you work with a contractor, he or she will be responsible for hiring and scheduling subcontractors; you will do this yourself if you are the contractor. To select subs, you can seek the advice of a contractor, ask your friends, or check the Yellow Pages or classified ads in your local newspaper. You may also want to visit a local construction site for a new home or a remodel and talk to the people working there.

When you have decided exactly what work you need to have done and when you want it completed, discuss your plans with the subs you have selected, show them your drawings, and get separate estimates for the cost of materials and labor. Then see what kind of a price you can get on the same materials from a cash-and-carry building materials supplier, and decide whether you want to supply your own materials or have the sub supply them. When you have decided which subs to employ, complete a contract form like the one in figure 4–1 and have it signed.

## GETTING ESTIMATES ON MATERIALS

If you order all your materials at the same time, you may be able to save a considerable amount of money and time. Many large-volume building materials suppliers have a sales representative who works specifically with owner-builders. This person's responsibility is to see that you receive good prices and to service your account.

Give your materials list to several suppliers and see where you can get the best price for the best materials. Do not be afraid to haggle—many suppliers have several pricing structures, and their salespeople can decide who gets what price. This usually holds true only for large-volume suppliers who provide materials for professional contractors, not cash-and-carry operations that cater to the do-it-yourselfers. A low price for inferior materials is no deal at all, however, so be sure you know what you are paying for. For example, the same size of lumber from two different suppliers can vary greatly in quality.

The following guidelines are intended to help you decide on the exact materials you want to use in your solarspace. Use these guidelines for pricing materials and getting estimates from subcontractors or building materials suppliers.

### General

- Use only standard-size building materials; avoid planning anything that requires custom-sized lumber, glass, or insulation and anything that requires special tools or added labor.
- Select materials that have surfaces and textures that harmonize with the materials used in the existing structure. For example, use the same kind of siding or veneer and roofing material on the solarspace as was used on the house.

- Select window and door units that are the same as or similar to existing windows and doors.
- Use locally manufactured materials whenever possible. For example, visit a sawmill, glass factory, adobe farm, or whatever is near you to see what you can buy to use in your solarspace.
- Use salvaged building materials if you want to save money, as well as to create a special look for your solarspace.

### Lumber

- Use construction-grade dimensional lumber if it will be covered with finish materials; use select structural-grade lumber when it will not be covered.
- Use fir, yellow pine, or whatever variety of wood that is strongest and least expensive in your area as framing lumber for your solarspace. If you can afford it, use cedar, redwood, or cypress, at least where the structural framing will be exposed, as in mullions and rafters that hold glazing units. (Use cedar for sill plates if possible.)
- Use redwood, cypress, northern cedar, yellow cedar (alderwood), or other weather-resistant wood for exterior or interior trim that will be exposed to the elements or moisture.

### Fasteners

- Use only galvanized nails and aluminum, brass, stainless steel, or galvanized screws or bolts that will not rust.
- If caulking and construction adhesives are used, select products that can withstand the temperature, humidity, and ultraviolet light conditions in a solarspace.

### Insulation

- Use standard sizes and thicknesses of fiberglass insulation material to insulate walls and ceilings whenever possible. If standard sizes will not fit between the studs and rafters of your solarspace, use expanded polystyrene beadboard, cut to fit snugly.
- Use 2 inches (5 cm) of Styrofoam type SM to insulate the outside of footings and foundation walls; apply with appropriate construction adhesive.

# SUBCONTRACT AGREEMENT

THIS AGREEMENT, made this_____day of_____19____, by and between_____ hereinafter called the Owner/Contractor, and_____ hereinafter called the Subcontractor.

For the consideration hereinafter named, the said Subcontractor covenants and agrees with said Owner/ Contractor, as follows:

**1.** The Subcontractor agrees to (furnish all material and) perform all work necessary to complete the

_____

_____

_____

for the above-named structure, according to the plans and specifications (details thereof to be furnished as needed) and to the full satisfaction of Owner.

**2.** The Subcontractor agrees to promptly begin said work as soon as notified by said Owner/Contractor, and to complete the work as follows:_____

_____

_____

_____

_____

**3.** The Subcontractor shall take out and pay for Workmen's Compensation and Public Liability Insurance, also Property Damage and all other necessary insurance, as required by the Owner/Contractor or by the State in which this work is performed.

**4.** The Subcontractor shall pay all Sales Taxes, Old Age Benefit and Unemployment Compensation Taxes upon the (material and) labor furnished under this contract, as required by the United States Government and the State in which this work is performed.

**5.** No extra work or changes under this contract will be recognized or paid for unless agreed to in writing before the work is done or the changes made.

**6.** This contract shall not be assigned by the Subcontractor without first obtaining permission in writing from the Owner/Contractor.

IN CONSIDERATION WHEREOF, the said Owner/Contractor agrees that he will pay to said Subcontractor the sum of ($_____)_____Dollars for above (materials and) work, to be paid as follows:_____percent (_____%) of all labor (and materials) that have been placed in position by said Subcontractor and balance_____

_____

_____

_____

until paid in full, after said Subcontractor has completed his work to the full satisfaction of all parties concerned, i.e.,    Owner/Contractor    Designer/Architect. If requested, Subcontractor agrees to furnish Owner/Contractor with Waiver of Lien for materials and labor.

The Owner/Contractor and the Subcontractor for themselves, their successors, executors, administrators and assigns, hereby agree to the full performance of the covenants of this agreement.

IN WITNESS WHEREOF, this agreement has been executed on the day and date written above.

_____        _____
          (Witness)                              (Owner/Contractor)

_____        _____
          (Witness)                              (Subcontractor)

*4–1. Subcontract agreement*

- Use foil-faced rigid insulation board such as Thermax, High-R, or Styrofoam TG for sheathing material on exterior walls if an exterior vapor barrier is desired. (Apply aluminized tape to joints.)
- Use a fiberglass (or other insulating) sill sealer between foundation and sill plate *or* lay the sill plate in a bed of caulking applied to the top of the foundation wall.

## Radiant/Vapor Barrier

- Use a continuous aluminized radiant/vapor barrier on the inside of the solarspace. Staple the barrier to the inside of studs and rafter, before the interior wall finish is applied, but after insulation has been added. Polyethylene sheeting is also frequently used as a vapor barrier in a solarspace. A ½-inch (1.3-cm) air space should be maintained between the vapor barrier and the interior wall covering.

## Sheetrock (Drywall; Gypboard)

- Use only water-resistant Sheetrock on the interior of a solarspace. Sheetrock can be nailed to studs and rafters or applied with a construction-grade adhesive and short concrete nails directly to the interior surface of a masonry wall as a substitute for plaster. Multiple layers of Sheetrock can be used to add mass to interior wall surfaces.

## Masonry

- Use poured concrete or filled or solid concrete block for foundation walls; figure out which would be least expensive in your area.
- Use dark (precolored or painted) "slump face" concrete block (solid or filled) for interior walls of the solarspace that will not be plastered or covered. (The rough surface of a slump-faced block—which should face the sun—has more surface area than a regular concrete block.)
- Consider using dense, dark-colored stone or brick as an interior wall covering to increase thermal mass in the solarspace.
- Use unglazed quarry tile, whole bricks or "pavers," stone, sun-dried clay tiles, or other dense materials as a covering for concrete floors; or use a specially formulated concrete stain or paint to color the floors. Solid 2-inch (5-cm) concrete cap block can also be used as a floor covering. These can be laid dry (without mortar) in any pattern desired, then covered with grout or thinset concrete and stained the desired color.

## Adobe

- If adobe is used in your area, consider building adobe end walls (insulated on the *out*side; see chapter 5) to increase thermal mass in your solarspace. Buy your adobe from a reputable dealer if you do not know how to make your own.

## FINANCING YOUR SOLARSPACE

Paying for your solarspace is like paying for anything else—either you have the money or you borrow it. Of course, the other option if you do not have all the cash you need is to save until you do.

### Saving

If you are willing to postpone the construction of your solarspace until you can save enough money to pay for it, you will be in a position to earn interest on your savings. Placing some of your money in high-yield certificates of deposit or mutual funds will help guarantee that you will not spend the money for something else in the meantime, and the interest payments will add substantially to your savings. If you have a regular savings account or certificates of deposit, it is sometimes possible to get a loan using your savings as security. Many institutions will make such a loan for as little as 2 percent above the rate of interest you are receiving on your savings account or certificates. So much for the good news; the bad news it that, while you are squirreling away your hard-earned cash, the prices of building materials, contracted labor, and conventional fuel are likely to increase and thereby erode the value of your savings.

# Borrowing

You can explore many options for borrowing money to build your solarspace. These include asking your Aunt Harriet, or another well-heeled relative, if she or he could "spare some cash to invest in solar futures," namely yours; or you may be able to borrow against the cash value of an insurance policy, remortgage your home, or take out a home improvement or personal loan from a bank, savings and loan, or finance company. Each of these options, except perhaps a loan from Aunt Harriet, should seriously be considered in relation to the interest rates, the term (repayment period), and the amount you could borrow.

The most important considerations to a potential lender reviewing your application for a home improvement loan are the amount of equity you have in your home, your apparent ability to repay the loan, and finally, the design of the specific improvements you have in mind. Because not all loan officers are familiar with solarspace additions, you may need to do some selling to convince them that the solarspace will increase the value of your property—both in terms of energy savings and added space. If you plan to do part or all of the construction, you may also need to convince the lender that you are a competent builder.

### Life Insurance

If you have owned a whole-life insurance policy for a number of years, it may have accumulated a sizable cash value. Depending on the terms of your specific policy and when it was written, you may be entitled to borrow an amount of money from the insurance company equal to the policy's cash value at a very low interest rate. For example, a policy written in 1965 may have an unbelievably low interest rate of 3 percent!

The great thing about borrowing on your life insurance is that you usually do not even need to tell the insurance agent why you want the money. Agents usually believe that the money is really yours, so if you say you want it, that is all they need to know. Chances are also good that you will get a check very quickly. Furthermore, you can have the option of paying only the interest on the loan, applying your annual dividends toward repayment of the principal. (Your insurance will be reduced by the amount of your loan balance, however, so if you die before your loan is repaid, your heirs will not get as much money.) Such a deal is hard to beat, so if you have a whole-life policy, pick up the phone and call your friendly insurance agent.

### Remortgaging

Remortgaging your home to finance a solar addition may be possible but is generally not a good idea, unless your original mortgage was written recently. If you have an open-ended mortgage and have accumulated a considerable amount of equity as a result of a sizable down payment, monthly payments, and appreciation of the property, the lending institution may be willing to rewrite the mortgage to finance your solarspace. The drawback to this deal is that the new mortgage will be subject to the current interest rates. Therefore, giving up a mortgage with a 9 percent interest rate, for example, and taking on a new mortgage at a 17 or 18 percent rate does not make much sense. The increased monthly payments resulting from the higher rate of interest alone would certainly more than cancel the energy savings from the solarspace. Futhermore, you could probably get a home improvement loan, using the equity in your house as collateral, that would make your total monthly payments less than they would be if you remortgaged your home to include the cost of the solarspace.

### Home Improvement Loans

Home improvement loans are usually available from banks or savings and loan associations, credit unions, and finance companies. If you are seeking such a loan, you should shop around for the best deal. Start with the institution that holds your mortgage if you have one. Since you have already established credit with that institution, you have an added advantage, provided, of course, that you have made regular payments. Furthermore, because the mortgage holder already "owns" your house in one sense (or a substantial part of it), the mortgaging institution may be more inclined to grant a loan to improve your (their!) property.

If you belong to a credit union, you may be

able to get a loan at a lower rate of interest than you could get from a commercial bank or savings and loan association. Because membership in a credit union is usually related to your place of employment and the credit union can deduct loan payments directly from your paycheck, they are often able to charge lower interest rates.

A home improvement loan from a finance company will probably carry the highest interest rate because such companies tend to make loans to less qualified borrowers. Apply for a loan from a bank or savings and loan first and borrow from a finance company only as a last resort. Some finance companies are also known to employ some rather severe penalties for late payments and in some cases for early payment (prepayment).

Before you sign on the dotted line for a home improvement loan, make sure you understand all of the provisions that pertain to it. For example, can you pay off the loan before it is due? Can you reduce the long-term interest charges by making advance payments on the principal? What are the provisions for late payments? The last thing you want to do is sign a loan contract that would cause you to lose your house if you became unemployed or disabled.

### Personal Loans

Other options besides using the equity in your home as collateral for a loan are available. If you own other property, such as unimproved land, an automobile, or a boat, you may be able to use the property as security (collateral) for a personal loan. The interest rates on such loans tend to be higher, and the loans are typically made for a shorter term than home improvement loans. Provided that you can handle the payments on a personal loan for a few years, the total cost of the money borrowed will be far less than it would be for a loan in the same amount spread over a longer period of time. If you explore all of the options, you should be able to secure a loan that is suited to your financial situation.

## STATE INCENTIVE PROGRAMS

Most states have some type of incentive program to encourage the use of renewable energy sources. These programs vary greatly from state to state and include such enticements as credits or rebates on state income tax, real estate or property tax relief, deferred appraisals of newly constructed solar structures, exemptions from state sales tax on materials used to build renewable energy equipment, and so on.

In some states the costs for materials and labor used to build a solarspace fully qualify for existing incentives, whereas in other states a solarspace does not qualify for incentive programs. If the incentives offered by your state are a determining factor in whether you will build a solarspace, check into the incentive programs before you get started. Certainly the more attractive such incentives as a tax credit or rebate are, the sooner you will recover the cost of your solarspace through energy savings *and* tax advantages.

## FEDERAL INCOME TAX CREDITS

The federal tax regulations in effect at the time of this writing allow a tax credit of up to $4,000 (40 percent of $10,000 of *qualified* expenses) for "materials and components whose sole purpose is to transmit or use solar radiation" (*Federal Register*, August 29, 1980). Credit may be claimed for expenses incurred between April 19, 1977, and January 1, 1986, for solar applications to your principal residence. (A tax credit is an amount subtracted from the total taxes you owe, so it is much more valuable than a deduction.) Furthermore, credits that exceed the tax you owe for a given year can be carried over to the subsequent years.

The regulations for energy credits are subject to some interpretation and may someday be revised to include all of the components used to construct solarspaces of the type described in this book. Unfortunately, at the present time, the costs of components that serve a dual function do not qualify for tax credits. For example, although the glazing of a solarspace serves as a solar collection surface, it also provides light and/or a view, and it encloses the space from the outdoors. The glazing would therefore be excluded from credits because it serves more than a sole function. The cost of components such as containerized phase change materials, which are used only for heat storage, should qualify, however, since the regulations state that "any shading, venting and heat distribution or storage systems that do not have a dual function will qualify." This part of the regulation would probably qualify shade cloth, nighttime insulation, roof vents, and fans used in a solarspace. The cost of building a masonry wall inside the solarspace would also be likely to qualify under these regulations if the masonry wall did not support any of the solarspace structure.

If you plan the features of your solarspace with the tax regulations in mind, you may be able to build your space so that more of the components would qualify for tax credits. For example, you might plan your solarspace as a "walk-in solar collector." This would require that the space be used for no other purpose than to collect solar radiation and that the heat produced in the solarspace be distributed, via fans, to the rest of the house.

Federal tax credits for solarspaces have been much debated. Some homeowners have successfully defended their right to receive tax credits for materials and components used in the solarspaces; others have not been so fortunate. The matter of what materials and labor costs will qualify for tax credits is still questionable, and the Internal Revenue Service continually makes rulings that affect the interpretation of tax legislation. Before constructing your solarspace, check the current regulations and recent rulings. Write: Internal Revenue Service, 1111 Constitution Avenue N.W., Washington, DC 20224, or call (202) 566–5000.

Whether or not you are able to take a full income tax credit on your federal return for your solarspace, you should have no difficulty receiving credit for the cost of adding insulation, caulking, weather stripping, storm doors and windows, and other energy-conservation devices used to upgrade the thermal efficiency of your house. (You may also be able to take credit for these kinds of items used in your solarspace if you build the solarspace first and add the energy-conserving features later.) The regulations provide for a credit of up to $300 (15 percent of $2,000 of *qualified* expenses) for energy-conserving materials and devices.

The regulations for energy tax credits are clearly stated in the IRS publication 903, "Residential Federal Energy Tax Credits"—pick up a copy at your local IRS office or write to the address above.

## WHENEVER YOU ARE READY

When you have completed planning your solarspace and have decided who will build it, when it will be built, and how to pay for it, you are ready to proceed with the question of how to build it. The next chapter contains step-by-step construction procedures and numerous illustrations that will help you construct your solarspace properly.

# FIVE
# Building Your Solarspace

## How should your solarspace be constructed?

Whether you will end up with a well-constructed energy-efficient solarspace is largely up to you. No matter who builds your solarspace, it will at least be necessary for you to understand the basic construction procedures so you can assure yourself that the work is proceeding as it should.

### BEFORE YOU BEGIN . . .

The items listed below will help you get your project underway and help to ensure that it will run smoothly thereafter. Complete these tasks before you begin construction of your solarspace:

1. Check legal liabilities, building codes, deed restrictions, and covenants if you have not already done so.
2. Secure a building permit if one is required.
3. Secure the start-up funds; sign loan papers or place your savings in an interest-bearing checking account.

4. Draw up a work schedule indicating who will do what, when (see example, fig. 5–1).
5. Sign legal agreements with contractor or subcontractors (see form, fig. 4–1).
6. Order the lumber and other building materials needed to build the solarspace (or have the contractor order them). If possible, arrange to have the materials delivered after the foundation work has been completed.
7. Order the glazing material if it will not be purchased locally. If the glazing is from a local supplier, have it delivered when you are actually ready to install it.
8. Say goodbye to peace and quiet around your house and prepare yourself for a period of organized chaos. Tell your friends you will be busy and not to come around for awhile unless they want to work.

### GETTING IT STARTED AND KEEPING IT GOING

The construction procedures described in this

# WORK SCHEDULE – WATKINS SOLARSPACE

| DATES | WORK TO BE COMPLETED | MATERIALS | WORKERS | COST SUPPLIES | / LABOR |
|-------|----------------------|-----------|---------|---------------|---------|
| MAR 19, 20 | LAY OUT FOUNDATION DIG FOOTING TRENCH POUR FOOTING | SHOVEL, PICK CEMENT MIXER, CEMENT | (OWNER) | $ 100. | –0– |
| MAR 26, 27 | BUILD STEM WALL INSTALL FOUNDATION BOLTS & VENTS | RECYCLED BLOCK BONDING COMPOUND BOLTS & VENTS | (OWNER) | 30. 36. | –0– –0– |
| APR 4–6 | FRAME IN SOLARSPACE INSTALL SHEATHING, SIDING, & ROOFING | LUMBER, SHEATHING, SIDING, & ROOFING | DUTCHMEN CONSTRUCTION | 755. | 650. |
| APR 7 | INSTALL GLAZING, DOOR, & WINDOW | 10 PATIO DOOR GLASS UNITS @ $60., DOOR, & WINDOW | HORNER GLASS | 600. 355. | 530. 100. |
| APR 7 | ROUGH IN ELECTRIC INSTALL FANS | WIRING, CONNECTORS, JUNCTION BOXES, & FANS | WATTS ELECTRIC | 170. | 80. |
| APR 8 | ROUGH IN PLUMBING AND COLLECTOR FINS | COPPER TUBING, FITTINGS, TANK, & COLLECTOR FINS | REAMER PLUMBING | 65. 65. 125. | 200. |
| APR 9, 10 | INSULATE WALLS & CEILING, INSTALL RADIANT/VAPOR BARRIER | INSULATION BARRIER & TAPE | (OWNER) | 87. 125. | –0– –0– |
| APR 16, 17 | FINISH CEILING TRIM-OUT INTERIOR | CEDAR SIDING FOR CEILING CEDAR TRIM | (OWNER) | 56. 34. | –0– –0– |
| APR 18 | BUILD INTERIOR STONE WALLS & FLOOR | STONE & MORTAR | SLICK'S ROCKWORKS | 96. | 120. |
| (SOMETIME BEFORE SUMMER) | INSTALL SHADE CLOTH | SHADE SCREENS FOR SOUTH WALL GLAZING SUNSCREEN FOR ROOF GLAZING TRACK | (OWNER) (OWNER) | 105. 60. 18. | –0– –0– –0– |
| (SOMETIME BEFORE WINTER) | MAKE & INSTALL RADIANT/INSULATING CURTAINS & SHADES | MATERIALS FOR PSDS CURTAIN & SHADES | (OWNER) | 250. | –0– |
| | | | | $3132. | $1680. |

5-1. Work schedule

TOTAL  $4812.00

chapter are written as step-by-step directions for the do-it-yourselfer. If you are not doing the actual work, use the information and illustrations to monitor the work others are doing for you. Bear in mind, however, that many ways can be used to accomplish the same thing, do not get into a hassle with people who are working for you about *how* they are doing a particular job, as long as they can produce the desired result.

To get an overview of the entire project, read through *all* the directions before you follow *any* of them. Then keep this book handy for reference when you actually get underway. If the directions appear to be written in a foreign language, study the illustrations carefully. If you still do not know a purlin from a penguin, ignore the directions and loan the book to the person who will build your solarspace.

# PREPARE THE FOUNDATION

A good foundation is crucial to the success of any building project—or anything else in life for that matter. Since the quality of the foundation will greatly affect the rest of your solarspace, make sure the foundation is done properly—no matter who does the actual work.

Before you start digging the foundation, carefully **inspect the area** where the solarspace will be built. After roughly determining where the outside walls of the space will be, check to see if the ground slopes away to the south, east, or west to permit adequate drainage and runoff. If it appears that water running off the roof or walls of the solarspace will be a problem, you may want to "rearrange" the earth a bit or plan to use spouting below the glazing surface to carry away (or even collect) rainwater. This is also a good time to plan for the drains inside the solarspace. Floor drains will be especially important if you are planning to grow food or include a hot tub or spa in the solarspace.

Be sure utility, telephone, or sewer lines do not run above or below the site; they could be inadvertently damaged during the construction process. Having a broken line repaired can be a very costly proposition, so check carefully. If you are uncertain about the location of underground gas, electric, or telephone lines, call the utility companies and ask them to check their records. Remember also that you may need overhead clearance for a backhoe and a cement truck around the entire foundation area. A fully loaded cement truck can weigh many tons, so check the route the truck will travel through your yard and plan to have it miss your underground septic tank, for example.

**Plan the type of foundation** you want to use. Before you go any further, make sure you know what the local building code requires and what type of soil your house rests on, then decide what type (or combination of types) of foundation will support your solarspace (see fig. 5–3). The foundation will extend around the entire east, west, and south walls of the solarspace. A good general plan would be to build the same type of foundation for your solarspace that your house has.

**Plan the exact size of the foundation** and the location of vents or doorways that will extend into the foundation wall. Be sure to allow the *exact* amount of space on the south wall of the solarspace to install the type, size, and amount of glazing you have selected. This is a very crucial decision, so plan the size of the foundation accordingly. (See page 85 for information on sill and mullion details.)

**Stake out the foundation** with string before you begin to dig. First drive a nail into the south wall of the house, about 6 inches (15.2 cm) above the ground (or drive a stake into the ground) where the outside surface of the east wall of the solarspace will connect to the house. Measure a length of string about 1 foot (0.3 m) longer than the width of the solarspace (from north to south) and tie one end of it to the nail. Then place one edge of a large framing square along the south wall of the house and align the string with the edge of the framing square that is perpendicular to the house. Drive a stake into the ground where the *outside* edge of the southeast corner of the solarspace will be and tie the other end of the string to it. Stake out the west wall in the same way, then tie a length of string between the stakes at the southeast and southwest corners of the solarspace. Now check the diagonal distance from northeast corner to the southwest corner of the foundation and from the northwest corner to the southeast corner. The diagonal distances should be the same in both directions; if they are, the foundation will be square—that is, it will have 90-degree corners and the south wall of the solarspace will be parallel to the south wall of the house.

*5-2. Ready, set, go! The Watkins residence, before solarspace (Photo: Darryl J. Strickler)*

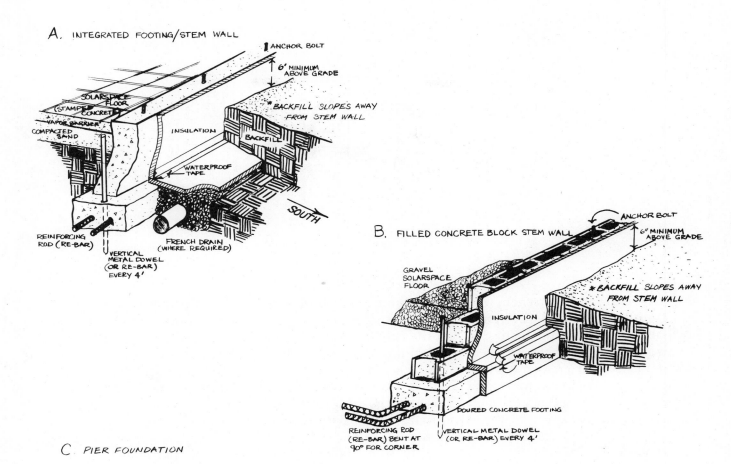

A. INTEGRATED FOOTING/STEM WALL

ANCHOR BOLT

6" MINIMUM ABOVE GRADE

SOLARSPACE FLOOR (STAMPED CONCRETE)

VAPOR BARRIER

COMPACTED SAND

INSULATION

* BACKFILL SLOPES AWAY FROM STEM WALL

BACKFILL

WATERPROOF TAPE

SOUTH

REINFORCING ROD (RE-BAR)

VERTICAL METAL DOWEL (OR RE-BAR) EVERY 4'

FRENCH DRAIN (WHERE REQUIRED)

B. FILLED CONCRETE BLOCK STEM WALL

ANCHOR BOLT

6" MINIMUM ABOVE GRADE

GRAVEL SOLARSPACE FLOOR

INSULATION

* BACKFILL SLOPES AWAY FROM STEM WALL

WATERPROOF TAPE

POURED CONCRETE FOOTING

REINFORCING ROD (RE-BAR) BENT AT 90° FOR CORNER

VERTICAL METAL DOWEL (OR RE-BAR) EVERY 4'

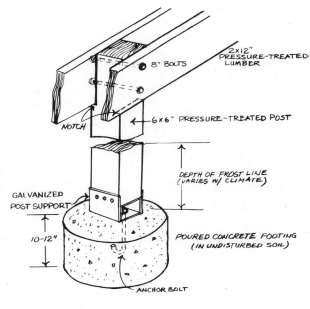

C. PIER FOUNDATION

8" BOLTS

2x12" PRESSURE-TREATED LUMBER

6x6" PRESSURE-TREATED POST

NOTCH

DEPTH OF FROST LINE (VARIES W/ CLIMATE)

GALVANIZED POST SUPPORT

10-12"

POURED CONCRETE FOOTING (IN UNDISTURBED SOIL)

ANCHOR BOLT

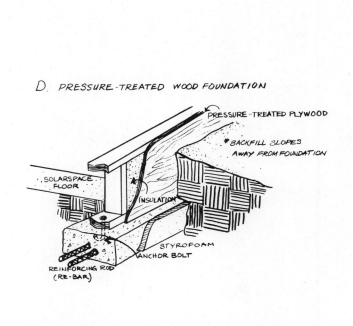

D. PRESSURE-TREATED WOOD FOUNDATION

PRESSURE-TREATED PLYWOOD

* BACKFILL SLOPES AWAY FROM FOUNDATION

SOLARSPACE FLOOR

INSULATION

STYROFOAM

ANCHOR BOLT

REINFORCING ROD (RE-BAR)

5-3. Types of foundations for solarspaces

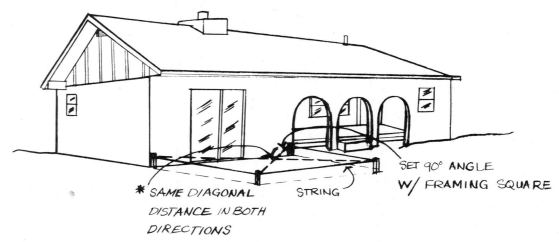

**SET 90° ANGLE W/ FRAMING SQUARE**

**STRING**

**\* SAME DIAGONAL DISTANCE IN BOTH DIRECTIONS**

5–4. *Lay out the foundation with string*

**Dig the footing trench,** an 18-inch-wide (45.7-cm) trench, 6 to 12 inches (15.2 to 30.4 cm) below the frost line. (The depth of the frost line varies with climate, so check the building standards in your area or ask at least two local builders how deep the footing should be.) To determine where the centerline of the footing should be, measure inside the string half the thickness of the stem wall (foundation wall). The center of the footing should be directly beneath the center of the stem wall.

5–5. *Footing trench; poured without forms (Photo: Darryl J. Strickler)*

**Build the forms** for the footing and the stem wall. If you can dig the footing trench (by hand or with a backhoe) so that it has vertical walls, you may be able to get by without using forms on the sides of the footing trench. If you cannot, use old lumber to build a form along the edge of the trench. Place 1½- or 2-inch (3.8- or 5-cm) Styrofoam along the inside of the south face of the trench, and lay two reinforcing rods (re-bar) about 8 inches (20.3 cm) apart along the entire length of the trench. (Put small rocks under the rods to hold the rods a few inches above the bottom of the trench or wait until you pour the concrete and simply drop the rods into the concrete while it is still wet.) Another alternative is to drive two metal dowels or re-bar into the ground (vertically) every 4 feet (1.2 m) and attach the two horizontal reinforcing rods to the vertical rods with wire.

If the stem wall will be constructed of concrete block, adobe, or pressure-treated plywood, you need only pour the footing. If, however, you want an integrated footing and stem wall, build a form similar to that in figure 5–6. Foundation insulation, 1½- or 2-inch-thick (3.8- or 5-cm) Styrofoam (*not* polystyrene beadboard) should be placed on the inside surface of the southernmost form wall, so it will adhere to the wall when the concrete cures. To do this, drive #16 galvanized nails through the insulation board so that the nail shafts protrude into the form as illustrated in figure 5–6; then temporarily attach the Styrofoam to the forms with double-headed nails. (If you do not want to insulate the stem wall at this point, plan to attach the insulation

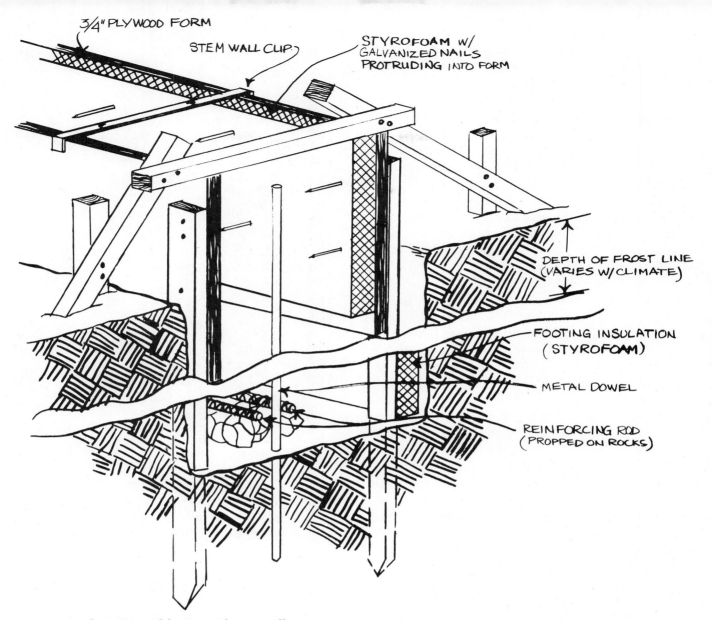

3/4" PLYWOOD FORM

STEM WALL CLIP

STYROFOAM w/
GALVANIZED NAILS
PROTRUDING INTO FORM

DEPTH OF FROST LINE
(VARIES w/CLIMATE)

FOOTING INSULATION
(STYROFOAM)

METAL DOWEL

REINFORCING ROD
(PROPPED ON ROCKS)

5-6. *Forms for integrated footing and stem wall*

board to the outside of stem wall with construction adhesive after the concrete cures and the forms are removed. The foundation may also be insulated on the inside, rather than the outside, if you find that more convenient.) Treat the exterior of the stem wall with waterproofing compound if it is customary to waterproof foundations in your locale.

As an alternative method of insulating the stem wall, specially designed polystyrene insulation such as Thermocurve can be integrated inside the stem wall when the concrete is poured. This kind of insulation eliminates the need to cover the insulation board—which is

necessary if the insulation is applied to the outside of the stem wall. A concrete block foundation wall can be insulated by pouring Zonolite into the hollow cores of the block.

Use a string chalkline to make a line on the inside of the form a few inches above where the top of the stem wall will be. Check the chalkline to be sure it is level. (It does not need to be parallel to the top of the form, but it must be level since it will serve as a guide when the stem wall is poured.)

If low vents, awning windows, or doorways are to be set into a poured stem wall, make frames for these openings with 2-inch-thick

(5-cm) lumber (treated with a wood preservative such as copper or zinc naphthenate) and temporarily attach the frames to the forms at the appropriate locations with double-headed nails. (When the concrete dries, these frames will, in effect, be bonded into the stem wall.) If you also drive galvanized nails through the frames so that the nails protrude into the cavity of the form, the frames will be anchored into the concrete when it dries. Vents or windows in a concrete block foundation wall can be integrated into the block course as the wall is constructed.

Finally, if part or all of the floor of the solarspace will be poured concrete, build the forms for the floor now.

**Pour the concrete** for the entire foundation at the same time. Depending on the size of the solarspace, the foundation wall and footing will require about 2 or 3 cubic yards (2.6 to 3.9 s) of concrete. (Forget about mixing the concrete yourself. Unless you are a real glutton for punishing exercise or you need very little, order it premixed.) Order enough concrete to pour the floors of the solarspace at the same time. (You may want to order the concrete with a water-proofing additive.) If the floor slab is poured up to the stem wall, place 1-inch (2.5-cm) rigid-board insulation (vertically) between the slab and the wall to allow for expansion of the concrete. While the concrete is still pliable, set anchor bolts (j-bolts) into the top of the stem wall every 18 inches (45.7 cm), within 4 inches (10.2 cm) of doorways and the corners of the stem wall. The threaded ends of the anchor bolts should protrude about 2 inches (5 cm) above the top surface of the stem wall. (Fill the top cores of a concrete block stem wall with grout mixture and set the anchor bolts in the filled cores. See fig. 5–7.)

If you will have poured concrete floors in your solarspace, consider adding a pebble or brush finish to the floor or a tile, brick, or stone pattern that is stamped into the floor while the concrete is still pliable (see page 101). After patterned concrete floors are stained and sealed, they closely resemble the real thing, at a fraction of the cost. You might also consider adding a coloring agent to the concrete so you will not need to stain or paint the floors later.

**Remove the forms** after the concrete has dried sufficiently. Depending on the outdoor temperature and humidity, this could take two

5–7. Recycled concrete block stem wall; filled with Zonolite insulation; top core grouted to accept anchor bolts; note opening for door (Photo: Darryl J. Strickler)

to five days. The warmer (and windier) it is, the sooner the concrete will set, although it takes as long as thirty days to harden completely.

## BUILD THE FRAMEWORK

Most of the structural elements of your solarspace will be constructed of standard-sized, dimensional framing lumber, such as $2 \times 6$s, $2 \times 8$s, or $2 \times 10$s. You might, however, consider building the east and west walls with filled or solid concrete block or with adobe (if available in your area) to add thermal storage mass to the interior of the solarspace. (Mass end walls need to be insulated on the outside, as described later in the chapter).

Do not reject the idea of building masonry end walls simply because you are not a skilled mason. You can either hire a mason to lay the block, or you can lay them yourself using a dry stacking method. This approach to building a wall is about as simple as it was in kindergarten to build a wall of wooden blocks: you simply stack the blocks dry, without mortar joints;

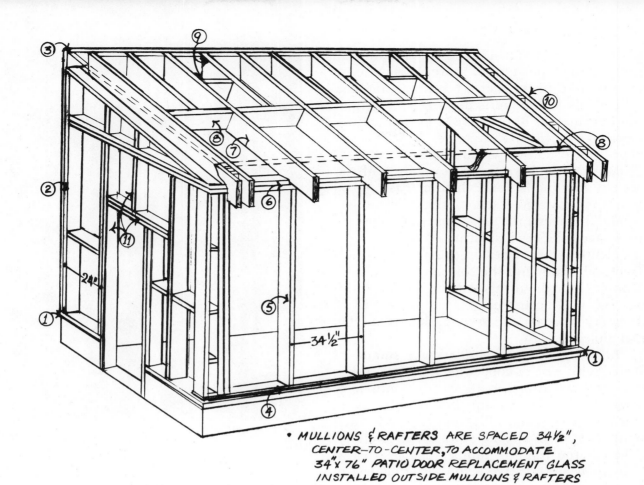

• MULLIONS & RAFTERS ARE SPACED 34½",
CENTER-TO-CENTER, TO ACCOMMODATE
34"x 76" PATIO DOOR REPLACEMENT GLASS
INSTALLED OUTSIDE MULLIONS & RAFTERS

FRAMING LUMBER — WATKINS SOLARSPACE (8'x 15')

| | | | | |
|---|---|---|---|---|
| ① | SILL PLATES | : 2-2x8x10' | CEDAR | $ 15.26 |
| | | 1-2x8x16' | " | 11.87 |
| ② | VERTICAL LEDGERS | : 2-2x6x10' | #3 FIR | 4.20 |
| ③ | HORIZONTAL LEDGER | : 1-2x10x16' | " | 6.72 |
| ④ | BOTTOM PLATE | : 1-1x6x16' | CLEAR FIR | 10.36 |
| ⑤ | MULLIONS | : 6-2x6x8' | " | 54.00 |
| ⑥ | TOP PLATES | : 2-2x6x16' | " | 36.00 |
| ⑦ | RAFTERS | : 6-2x8x12' | " | 108.00 |
| ⑧ | BLOCKING | : 2-2x8x12' | " | 36.00 |
| | | 1-2x8x6' | " | 9.00 |
| ⑨ | HALF RAFTERS & FRAME FOR VENTS | : 3-2x8x10' | #3 FIR | 8.46 |
| ⑩ | OUTSIDE RAFTERS | : 2-2x8x12' | " | 7.81 |
| ⑪ | STUDS, JAMBS, & PURLINS FOR END WALLS | 30-2x6x8' | " | 50.40 |

TOTAL - $358.08 *

* ACTUAL PURCHASE PRICE -6/82 w/10% BUILDERS' DISCOUNT
{IF #3 FIR (CONSTRUCTION GRADE) WERE USED INSTEAD OF "CLEAR FIR"
("C" GRADE - SELECT STRUCTURAL) THE TOTAL COST WOULD BE $149.55}

5-8. Structural framing of Watkins solarspace

then cover both the inside and outside wall surfaces with a plaster-type mixture of concrete and fiberglasslike fibers such as Surewall or Maxbond. (This technique is described in Agriculture Information Bulletin #374 available from: Superintendant of Documents, U.S. Government Printing Office, Washington, DC 20402.)

Masonry end walls are normally built before any framing is done. In some cases, however, it is possible—and sometimes desirable—to build them after the roof and south wall of the solarspace have been constructed.

Figure 5–8 provides an overall view of the structural framing of the solarspace used as an example throughout this chapter. Although this example will be useful as a guide, it will be necessary to adapt the specific details to your own project.

All wood used in a solarspace should be treated with zinc or copper naphthenate. When dry, these wood preservatives are nontoxic to plants, but in liquid form they have an awful odor and are very unhealthy for people and other living things, so use and store them carefully. Although the framing lumber can be treated with wood preservative before or after it is in place, it is often easier and more effective to apply the preservative to the raw lumber. (Pressure-treated wood such as Wolmanized lumber and cedar do not need to be treated with a wood preservative.)

**Lay the sill sealer,** a ½-inch-thick (1.3-cm) layer of fiberglass or other insulating or sealing material, along the top edge of the stem wall to reduce air infiltration between the sill plate and the stem wall. Sill sealer insulation usually comes in 3½- or 6-inch (8.9-cm or 15.2-cm) widths, so you may need to place two or more widths of it along the entire length to completely cover the top surface of the stem wall. As an alternative to sill sealer insulation material, which can become waterlogged with moisture, put down a bedding of Butyl caulk or mastic between the underside of the sill plate and the top surface of the stem wall.

**Install sill plates,** 2-inch-thick (5-cm) dimension lumber that is as wide as the combined width of the stem wall and the rigid-board insulation on the outside of the stem wall. Use a single piece of lumber for the sill plate on the south wall. Butt, miter, or cut an overlapping joint for the corners where the sill plates for the

5-9. *Place sill sealer on the top of the stem wall before installing the sill plate (Photo: Darryl J. Strickler)*

east and west walls meet the sill plate on the south wall. Unless you use cedar for sill plates—which would be a good idea—treat them with zinc or copper naphthenate. When the wood preservative has dried, lay the sill plates along the top of the stem wall and mark the exact location of the anchor bolts protruding from the stem wall. Then drill the appropriate-size holes through the sill plates and bolt them in place. Countersink the bolt heads.

**Install ledger plates** where the roof and the end walls of the solarspace will attach to the house. Ledgers are usually 2-inch (5-cm) dimensional lumber, either 2 × 6s, 2 × 8s or 2 × 10s—depending on what size is being used to frame the roof and walls of the solarspace. Ledgers should be attached to the wall of the house with ¼-inch (0.6-cm) lag screws or expansion bolts 6 inches (15.2 cm) long. Lag screws are screwed directly into the studs in a frame wall; expansion bolts are screwed into lead anchors (expansion shields) that have been drilled into a masonry wall.

If the surface of the wall is uneven, as when it is covered with siding or brick—staple sill

5-10. *Run a bead of caulking along sill sealer; mark the exact location of anchor bolts; then drill holes for anchor bolts (Photo: Darryl J. Strickler)*

sealer insulation to the back of the ledgers before fastening them to the wall. If the wall is stone with a very uneven surface, plaster the area of the wall where the ledgers will be attached with a coat of thinset concrete to prepare an even surface on which to attach the ledgers.

To install the *horizontal ledger* where the roof of the solarspace ties into the house, follow the directions below that apply to your particular situation.

If the roof ridge line of the house runs east and west and

■ If the existing overhang of the south roof of the house will be the only solid (insulated) portion of the roof of your solarspace, remove the fascia board and spouting (if any) from the overhang and bolt the horizontal ledger directly to the ends of the house roof rafters from which you removed the fascia board. If the fascia board is in good condition, leave it in place and nail a 1-inch-thick (2.5-cm) board of the same width as the fascia board over it.

■ If the solid portion of the solarspace roof is to be a continuation of the existing south-sloping portion of the house roof, cut back the rafter tails and the soffit (overhang) to

5-11. *Cedar sill plate being installed over anchor bolts (Photo: Darryl J. Strickler)*

5-12. *Staple sill sealer to the back of ledger plates (Photo: Darryl J. Strickler)*

5-13. *Plaster uneven walls with thinset concrete before installing vertical ledger plates (Photo: Darryl J. Strickler)*

■ If the pitch of the solarspace roof is to be, for example, 3:12 (that is, if the roof rises a vertical distance of 3 feet in a horizontal distance of 12 feet) and the pitch of the existing south-sloping portion of the house roof is 5:12—and if the roof of the solarspace must be tied in higher on the south roof of the house than where it meets the south wall—cut back the soffit over the south wall. Then remove the courses of shingles from the house roof to the point where the solarspace roof and house roof will meet, attach the ledger plate directly over the roof decking, and screw it into the existing rafters (see fig. 5–14).

If the roof ridge line of the house runs north and south and

■ If you are building a solarspace on the south end wall of the house, attach the horizontal ledger at the appropriate height by nailing it (temporarily) in place. When you are certain that the ledger is level, remove one nail at a time and replace it with a lag screw that is screwed into the studs inside the wall (or into lead expansion shields set into a masonry wall).

■ If the roof of the solarspace is to be tied into the gable of the south wall of the house, attach ledger(s) in the appropriate locations.

If the house has a flat roof, attach the horizontal ledger for the solarspace directly to the south wall of the house.

To install *vertical ledgers* where the east and west walls of the solarspace meet the south wall of the house, follow the directions below that apply to your particular situation.

the point where they meet the south wall of the house. Then attach the horizontal ledger plate to the shortened rafter tails and the top plate on the south wall of the house.

■ If the roof of the solarspace is to be tucked under the existing south-sloping portion of the roof of the house, attach the horizontal ledger directly to the south wall of the house. Be sure to allow enough space under the existing soffit to apply flashing, decking, and shingles for the roof of the solarspace.

5–14. *Horizontal ledger plate installed on roof when south wall is too low to install ledger on wall*

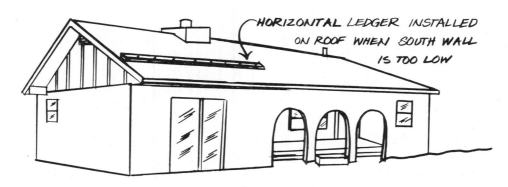

HORIZONTAL LEDGER INSTALLED
ON ROOF WHEN SOUTH WALL
IS TOO LOW

If the east and west walls of the solarspace are masonry, no vertical ledgers are required. The masonry walls of the solarspace should, however, be tied into the south wall of the house with brick ties.

If the east and west walls of the solarspace are wood-frame construction:

1. Use a plumb bob and level to set the vertical ledgers. Align the outside edge of the ledger plate to the outside of the sill plate.
2. Attach the ledgers to the south wall of the house with lag screws. Screws should meet a stud in a frame wall or a lead shield drilled into a masonry wall. You may have to install blocking or another stud, face out, in the wall if existing studs do not line up with the ledgers.
3. Toenail the vertical and horizontal ledgers, and the vertical ledgers and the sill plate, together—that is, drive nails at an angle across the corners.

**Decide which sill and mullion details** you want to use before you proceed any further. The details in fig. 5–15 are designed to prevent leakage and provide adequate support for the glazing. Note that glazing can be installed either outside the mullions or between them. The specific glazing and sill details you use will affect the distance between mullions, center to center, and how far the mullions are set back from the outer edge of the sill plate. Please note that patio door glass replacement units must have at least ¼-inch (0.6-cm) clearance around all edges to allow for expansion and prevent breakage of the seals.

If you (or someone else) decides to devise your own sill and mullion details, be sure the construction you use will prevent potential problems caused by air or water leakage and will provide equal support to both panes of glass along the bottom edge of double-insulated units.

**Choose your roof/wall details** from those illustrated in figure 5–16. Study each possibility carefully and decide which details, or combination of details, is best suited to your particular project, your sense of aesthetics, and your carpentry skills. Remember that the last rafter and mullion at each end of the roof and wall is to be double width; use either 4-inch-thick (10.2-cm) lumber or nail two 2-inch (5-cm) pieces together with ⅜-inch-thick (1 cm) exterior-grade plywood between them.

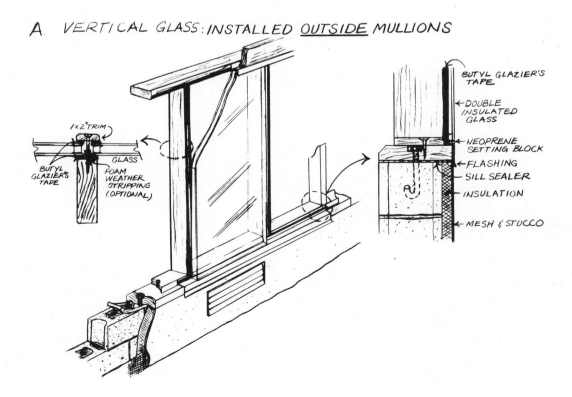

A VERTICAL GLASS: INSTALLED OUTSIDE MULLIONS

5–15. Sill and mullion details

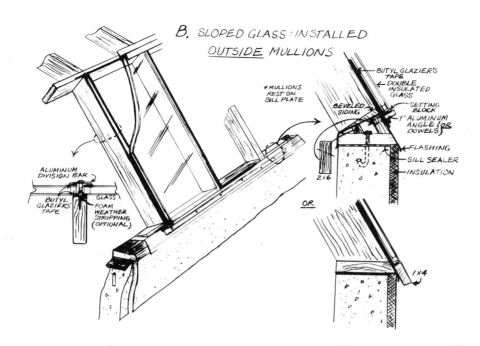

B. SLOPED GLASS: INSTALLED __OUTSIDE__ MULLIONS

* MULLIONS REST ON SILL PLATE

ALUMINUM DIVISION BAR

BUTYL GLAZIERS TAPE

GLASS FOAM WEATHER STRIPPING (OPTIONAL)

BUTYL GLAZIER'S TAPE

DOUBLE INSULATED GLASS

BEVELED SIDING

SETTING BLOCK

1" ALUMINUM ANGLE (OR DOWELS)

FLASHING

SILL SEALER

INSULATION

2×6

__OR__

1×4

C. VERTICAL GLASS: INSTALLED __BETWEEN__ MULLIONS

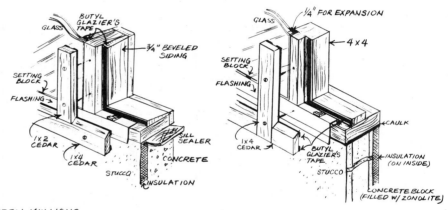

GLASS

BUTYL GLAZIER'S TAPE

¾" BEVELED SIDING

SETTING BLOCK

FLASHING

1×2 CEDAR

1×4 CEDAR

SILL SEALER

CONCRETE

STUCCO

INSULATION

GLASS

¼" FOR EXPANSION

4×4

SETTING BLOCK

FLASHING

1×4 CEDAR

BUTYL GLAZIER'S TAPE

CAULK

INSULATION (ON INSIDE)

STUCCO

CONCRETE BLOCK (FILLED W/ ZONOLITE)

D. SLOPED GLASS: INSTALLED __BETWEEN__ MULLIONS

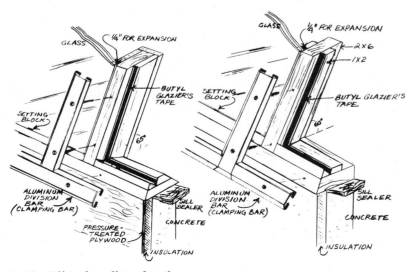

GLASS

¼" FOR EXPANSION

BUTYL GLAZIER'S TAPE

SETTING BLOCK

65°

ALUMINUM DIVISION BAR (CLAMPING BAR)

SILL SEALER

CONCRETE

PRESSURE-TREATED PLYWOOD

INSULATION

GLASS

¼" FOR EXPANSION

2×6

1×2

SETTING BLOCK

BUTYL GLAZIER'S TAPE

ALUMINUM DIVISION BAR (CLAMPING BAR)

SILL SEALER

CONCRETE

INSULATION

5-15. *Sill and mullion details*

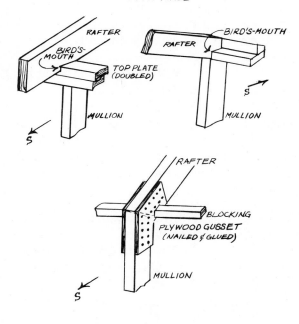

VERTICAL SOUTH WALL

RAFTER

BIRD'S-MOUTH

BIRD'S-MOUTH

TOP PLATE (DOUBLED)

MULLION

MULLION

S

RAFTER

BLOCKING

PLYWOOD GUSSET (NAILED & GLUED)

MULLION

S

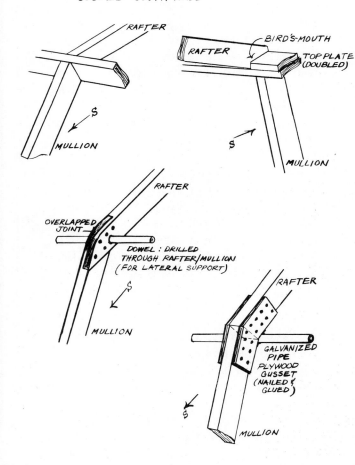

SLOPED SOUTH WALL

RAFTER

BIRD'S-MOUTH

RAFTER

TOP PLATE (DOUBLED)

MULLION

S

MULLION

S

RAFTER

OVERLAPPED JOINT

DOWEL: DRILLED THROUGH RAFTER/MULLION (FOR LATERAL SUPPORT)

S

MULLION

RAFTER

GALVANIZED PIPE

PLYWOOD GUSSET (NAILED & GLUED)

S

MULLION

*5-16. Roof-wall details*

**Frame the south wall and roof** of the solarspace. Select the directions from those given below that apply to your particular situation.

1. If you use a *bird's-mouth* roof/wall detail similar to those illustrated in figure 5–16, you may find it easier to prefabricate the south wall of the solarspace on the ground; then stand it in place and prop it temporarily with bracing until you have built the roof rafters that will hold it permanently in place. After the south wall is in place, nail (screw or bolt) it to the sill plate. Next, cut one rafter to length and notch it; if it fits perfectly, use it as a pattern for the rest of the rafters. Cut all the rafters exactly the same and nail them to the top plate on the south wall of the solarspace and to the ledger plate on the south wall or roof of the house (see step 3 below).

2. If you use a *butt and gusset* roof/wall detail, it will be easier to prefabricate each combined rafter/mullion separately. First build one to use as a pattern; if it fits perfectly, make the rest identical. After all of the rafter/mullions are placed exactly where they belong and are nailed permanently to the ledger and the sill plate, brace them (temporarily) with a board tacked across the south face of the framing (see step 3 below).

   Although the plywood decking on the solid (insulated) portion of the solarspace roof and the glazing units themselves will usually provide adequate lateral support, you may want to add permanent bracing (blocking) between the rafter/mullions to add strength to the structure. Blocking of the same-size lumber as the rafter/mullions should be nailed between the rafter/mullions (see page 89). A length of galvanized pipe that runs *through* all of the rafter/mullions or is strapped to the back of each one can serve as lateral bracing (and a handy place to hang potted plants).

   If you are going to use fiberglass or another type of roll glazing, you may want to add purlins (blocking) between the mullions on the south-wall glazing area. Purlins should be placed where they can serve as a base to join overlapping edges of the glazing.

5-17. *South wall of solarspace being framed on ground (Photo: Darryl J. Strickler)*

5-18. *South wall of solarspace propped in place and braced (Photo: Darryl J. Strickler)*

5-19. *South wall braced; rafters installed (Photo: Darryl J. Strickler)*

3. Use joist hangers to attach the rafters to the horizontal ledger plate on the south wall (or roof overhang) of the house or toenail the rafters to the ledger and add blocking between the rafters as shown in figure 5-20.
4. Toenail the ends of the mullions to the sill plate with #16 galvanized finish nails.

**Frame the east and west end walls** unless they are constructed of masonry and adobe. Use 2 × 6 framing lumber for end walls to allow more space for insulation in the unglazed portion of the walls.

1. If the south wall of the solarspace is sloped, place a stud (vertically) directly beneath the intersection of the roof and double-end rafters and the top plate of the south face of the solarspace (see fig. 5-8).
2. Place a stud (horizontally) between the intersection of the roof end rafters and the header plates of the south face of the solarspace and the vertical ledger plate on the south wall of the house. Toenail one end of this horizontal stud to the top plate and the other end to the vertical ledger.
3. Frame the rest of the wall; rough in openings for any doors, windows, or vents. Studs should be placed 16 or 24 inches (40.6 or 60.9 cm) center to center, to accommodate standard-size fiberglass insulation. Use the same details that you used in the south wall for any glazing in the east and west walls. You may, however, want to use different-size glazing units than you used in the south wall, or even operable windows.
4. Run wiring to any openings in the walls that will have electrically operated fans or a heat exchanger and to locations of light switches, electrical outlets, or lighting fixtures.

This is a good time to treat the wooden framework of the solarspace with zinc or copper naphthenate if you did not pretreat the lumber. It is also an ideal time to paint any framing that will ultimately show with a highly reflective white paint or to stain it with an exterior-grade stain if you prefer. There is no need to paint or stain the framework if it will be covered with finish lumber or other materials. All of the framing must, however, be treated with a wood preservative.

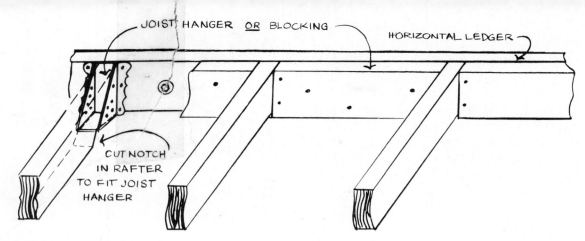

5-20. *Joist hangers and blocking*

## ENCLOSE THE ROOF

If the roof of the solarspace will be partially glazed, decide what portion of the roof will be glazed in relation to the type and size of roof glazing you select. Translucent glazing works best in the roof area of a solarspace because it spreads the light more evenly through the space. Because the glazed area of the roof will be subject to the extreme heat losses, you might consider installing tempered, double-insulated glass between the rafters as the outer glazing and a layer of light-diffusing glazing to the underside of the rafters as an inner glazing. (This kind of setup requires that weep holes be cut to prevent condensation from building up in the space between the inner and outer glazing surfaces.) A low-cost, easy-to-install alternative for roof glazing is corrugated fiberglass glazing installed over the top of the roof rafters. Whether roof glazing is installed between the rafters or on top of them depends on how you plan to install the glazing in the south wall of the solarspace. For example, if the south wall glazing is to be installed outside the mullions, it would probably work best to install the roof glazing on top of the rafters.

To enclose the roof of your solarspace, follow the directions below that apply to your particular situation.

If the roof of the solarspace will be partially glazed:

1. Mark the tops of the rafters to indicate the dividing line between the glazed and unglazed portion of the roof. (Use a string chalkline or measure the exact length on each rafter.) Cut pieces of dimensional lumber (of the same size as the rafters, as long as the distance between rafters) to serve as blocking between the rafters. Use the marks on the top of the rafters to align the blocking in a straight row across the roof (see fig. 5–21). Toenail the blocking in place. If additional blocking is needed in other parts of the roof area, it may be staggered or offset, which will allow you to nail the blocking through the rafters rather than toenailing it in place.

2. Build the glazing stops for the roof glazing if it is to be installed *between* the rafters. For ease of installation, corrugated fiberglass or flat, translucent glazing material can be nailed to the *top* of the rafters with rubber gasketed nails. If corrugated material is used, a 1- by 2-inch (2.5- by 5-cm) board cut to match the corrugation must be placed under the front and back edge of the glazed portion of the roof, and a half-round molding strip (the same diameter as the width of the corrugations) should be tacked to the top surface of each rafter.

   At this point you should prepare the roof so that the glazing can be installed, but do not actually install it until the rest of the roof is completed.

If part or all of the roof area is to be insulated:

1. Install blocking between the rafters (see step 1, above) and rough in openings in the roof for any operable skylights, roof windows, power roof vents, or wind turbines you plan to use.

2. Install roof decking, ½- or ¾-inch (1.3- or 1.9-cm) exterior-grade plywood nailed to the top of the rafters. The shingles should slightly overhang the glazed portion of the roof, so if the roof glazing is installed on top of the rafters, you may need to build up the height of the roof surface with rigid insulation board or insulated sheathing between the top of the rafters and the roof decking. The roof decking must also overhang the side walls enough to cover the top edge of insulated sheathing (if used) and siding (or other veneer) on the walls. Cut out openings in the roof decking for skylights, windows, or vents. Corrugated metal roofs and cedar shake shingles do not require plywood decking or roofing felt. Shakes are nailed to 1- by 4-inch (2.5- by 10.2-cm) strips, 4 inches (10.2 cm) apart; corrugated metal roofing is installed directly on top of the rafters in the same manner as corrugated fiberglass is installed (see step 2, page 89).

3. Apply roofing felt (tar paper) or PARSEC Airtight-White horizontally over the decking. Start at the bottom of the roof, overlap the seams of successive runs, and staple the felt to the roof at the overlapped seams with a staple gun. If you shingle-as-you-go, you do not actually need to fasten the roofing felt since the nails through the shingles will hold the felt in place.

4. Install flashing where the roof of the solarspace meets the house, over the end rafters and down the side walls, around openings in the roof, under the shingles that will overhang the glazed portion of the roof, *and* anywhere else the roof is apt to leak. Set the flashing in a bed of roofing cement.

5. Install shingles or other roofing material to match the roof of the house. Shingles should overhang the plywood roof decking by about ½ inch (1.3 cm). A drip edge (flashing) may be installed under the starter course of shingles. Shingles applied to a roof with a pitch as low as 3:12 should have no more than a 4-inch (10.2-cm) exposure per course.

If you are working during cool weather or if there is a possibility of rain, temporarily cover the glazing areas of the solarspace with polyethylene sheeting at this point.

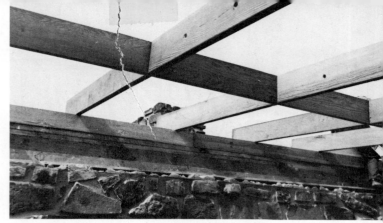

5-21. *Blocking between rafters marks dividing line between glazed and unglazed portion of the roof; holes through rafters will accommodate collector pipes for solar hot water system (Photo: Darryl J. Strickler)*

5-22. *Plywood decking of roof; open area for roof glazing (Photo: Darryl J. Strickler)*

5-23. *Insulated sheathing applied under plywood decking; hole cut for wind turbine; wiring for fans and lights roughed in (Photo: Darryl J. Strickler)*

# ENCLOSE THE END WALLS

The exterior portion of the end walls that will not have glazing should first be covered with an insulated sheathing or rigid insulation board. Polystyrene or Styrofoam insulation board may be installed on the outside of the studs and extend down over the outside of the stem wall, all the way to the footing. If you live in an area where an exterior vapor barrier is desirable, use foil-faced rigid-board insulation such as Thermax as sheathing or add a separate layer of aluminized film over the sheathing. A foil layer under the siding will serve as a radiant, infiltration, and vapor barrier on the outside of the wall. Du Pont Tyvek or PARSEC Airtight-White may be used instead of foil as an air infiltration barrier. Siding (or other veneer) to match the house should be installed on the outer surface of the wall; fascia and trim boards, installed under the roof overhang at the top of the end walls. Fascia and trim should match that used on the house.

Follow the directions below that apply to your particular situation.

**Nail insulating sheathing to the outside of the studs** with special nails that have a square washer under the nail head (cap nails). Do not cover door, window, or vent openings. Tape all of the joints in the sheathing with aluminized pressure-sensitive tape to reduce air infiltration. (If the sheathing has a foil layer, the foil side should face *out*. If the sheathing does not have a foil layer and an exterior vapor barrier is desirable, apply an aluminized film such as PARSEC Thermo-Brite on the outside of the sheathing, unless you plan to stucco the outside of the wall.) Tape all joints in the foil layer with aluminized tape. Install PARSEC Airtight-White (Du Pont Tyvek) instead of foil on the outside of the sheathing (as an air-infiltration barrier) if you do not need an exterior vapor barrier in your locale. Tape all joints with aluminized tape (*not* duct tape).

**Install L flashing** vertically where the side walls of the solarspace meet the wall of the house. Lay the flashing in a bed of waterproofing compound and nail it to the studs at the end of the wall. Nail through the sheathing and apply waterproofing compound over the nail heads, or use gasketed nails.

**Apply siding or other veneer** to the outside of the wall, over the sheathing. The insulating

5-24. *Roofing felt and shingles being applied; wind turbines integrated into roof (Photo: Darryl J. Strickler)*

value of foil-faced sheathing or aluminized film on the outside of the sheathing will be increased if the siding or veneer is spaced (furred out) ½ to ¾ inch (1.3 to 1.9 cm) from the foil, thus creating a dead air space. A brick veneer can be laid on the footing and extend up the wall to the desired height. A stucco-type finish using a system such as Dryvit or Settef can be applied directly to the Styrofoam sheathing on the end walls and stem wall if desired.

**Install fascia and trim** boards at the top of the walls where the walls meet the edge of the roof.

5-25. *Exterior wall covered with insulated sheathing; joints taped; fascia and trim board installed (Photo: Darryl J. Strickler)*

# INSULATE THE ROOF AND WALLS

Fiberglass batts are the least expensive, most readily available type of insulating material for roofs and walls. Standard-size batts are designed to fit between studs or rafters placed 16 or 24 inches (40.6 or 60.9 cm) apart, center to center, and are available in 3½-, 6-, and 9-inch (8.9-, 15.2-, and 22.9-cm) thicknesses with insulating values of R-11, R-19, and R-30 respectively.

If your roof rafters or wall studs are not 16 or 24 inches (40.6 or 60.9 cm) apart, you might consider cutting and piecing standard batts together or using expanded polystyrene beadboard (which usually comes in your basic white) of the appropriate thickness and width. Polystyrene has more insulating value per inch than fiberglass and is not affected the way fiberglass is by moisture. Standard-size beadboard is 1, 1½, 2, 3, 4, 6 or 8 inches (2.5, 3.8, 5, 7.6, 10.2, or 20.3 cm) in thickness, available in 2- by 8-foot or 4- by 8-foot (0.6- by 2.4-m or 1.2- by 2.4-m) sheets. Insulation should completely fill the space between the studs and rafters. This is a slight problem since 2 x 6 studs are actually 5½ inches (14 cm) wide and 2 x 8 rafters are 7½ inches (19 cm) wide. Although it can be easily cut (more like frayed) with a saw, the best way to cut polystyrene is with a "hot-wire." See if you can have the supplier cut the polystyrene to custom-fit the size you need—it will fit much tighter if it is cut with a hot wire.

Fiberglass blankets, usually also without vapor barriers, are sized to be press-fitted between studs and rafters. You may want to use these instead of batts.

Although fiberglass insulating batts usually have a treated paper covering, do not count on the covering to serve as an adequate vapor barrier. Because of the unusually high moisture content in the air of a solarspace filled with plants, you will need an absolutely airtight continuous vapor barrier on the interior to keep moisture out of the walls and ceiling. (Moisture normally travels from inside to outside.) Although polyethylene sheeting is often used as a vapor barrier, in the high moisture conditions often found in a solarspace, water can condense between the polyethylene and the interior finish and can cause problems like falling Sheetrock

and rusted-out nails. Moreover, polyethylene serves as only a vapor barrier; a single radiant/vapor barrier, serving a dual function, is more efficient. The most effective vapor barrier that also doubles as a radiant barrier is aluminized film. A radiant/vapor barrier works best if a ¾-inch (1.9-cm) air space is left between it and the wall or ceiling covering.

A newer type of insulation, Foilpleat, appears to have considerable potential—especially for use in solarspaces. It consists of multiple layers of aluminized film with kraft paper backing and has dead air spaces between layers. Thus, the insulation can serve as a radiant/vapor barrier as well as reduce heat flow.

Masonry walls need to be insulated on the *outside*. The easiest way to do this is to apply polystyrene beadboard to the outside of the walls and cover it with a stucco finish. Other techniques for finishing the outside of a masonry wall are described below.

Select the directions below that apply to your particular situation.

**Install insulation between studs and rafters** on the portions of the roof and walls that will not be glazed. If you are using fiberglass batts, staple the paper flanges directly to the studs or rafters; overlap the flanges on adjoining batts. Be sure adjoining batts between the same two studs or rafters are tightly butted together. If you are installing fiberglass blankets or polystyrene beadboard, these materials should be cut to a size that can be press-fitted so the material will stay in place until the vapor barrier and interior finish material are installed to hold it permanently. (Follow manufacturer's instructions carefully when installing Foilpleat insulation. Air pockets between layers and distance from exterior and interior wall finishes will affect the efficiency of Foilpleat.)

Be kind to yourself—wear long pants, a long-sleeved shirt, a hat (perhaps covered with mosquito netting), gloves, and a dust mask when working with fiberglass insulation. Your skin and lungs will thank you. (These precautions do not apply to the installation of Foilpleat.)

**Install a radiant/vapor barrier** such as PARSEC Thermosol Brite on the interior of the solarspace. This barrier should be continuous and completely cover insulated roof and wall surfaces (including the header and sill plates) without a break where the roof and wall surfaces meet. Staple the barrier to the studs and rafters at 3-inch (7.6-cm) intervals; then use aluminized pressure-sensitive tape to cover the staples and seal the seams between sheets. Also tape around windows, doors, and vents. Try to achieve a completely airtight seal so that moisture cannot condense inside the walls and ceiling. Allow a ¾-inch (1.9-cm) air space between the radiant barrier or Foilpleat insulation and the material covering the wall on the inside.

The techniques described below for insulating the outside of masonry walls will add 5 inches (12.7 cm) or more to the thickness of the finished wall. Be sure to compensate for this added wall thickness (and fascia and trim boards) when you apply the roof decking.

**Insulate masonry or adobe end walls** by attaching 4-inch (10.2-cm) expanded polystyrene beadboard (or four layers of 1-inch [2.5-cm] rigid insulation board, spaced ½ inch [1.3 cm] apart) directly to the outside of the wall with specially formulated Styrofoam adhesive. You can also use stick-clips—long nails with a square, perforated head on one end that is glued to the wall and a washerlike device on the other end—to attach the insulation. The beadboard may run all the way down to the footing, but only the aboveground structure must be covered. The exterior surface of the Styrofoam or beadboard may be finished with a commercially available mesh

and stucco finish such as Dryvit or Settef, or chicken wire may be stapled over the insulation board and a home-mixed exterior-type stucco applied over it. You can also apply exterior-grade plywood siding directly to the beadboard with construction adhesive or apply a brick or masonry veneer as the outside surface of the wall. If you use a masonry veneer, apply PARSEC Thermo-Brite adhesive-backed aluminized film to the outside of the beadboard and leave a ¾-inch (1.9-cm) dead air space between the wall and the veneer. If you need a really low-cost temporary solution, simply paint the beadboard with a good exterior-grade latex paint.

**To cover the outside of masonry walls with siding,** first apply an adhesive-backed aluminized film to the outside surface of the masonry wall. Then build a 2 × 4 stud-frame wall, ¾ inch (1.9 cm) away from the outside surface of the masonry wall. (You may need to pour a separate footing for the stud-frame wall on top of the existing stem wall footing.) Insulate the spaces between the studs with 3½-inch (8.9-cm) fiberglass or 4-inch (10.2-cm) polystyrene beadboard cut to fit between the studs. Apply an insulated sheathing material to the outside of the

5–27. PARSEC Thermo-Brite used as radiant/vapor barrier on interior of solarspace (Photo: Darryl J. Strickler)

5–26. Polystyrene beadboard custom-cut to fit 7½-inch thickness between 2x8 rafters of solarspace (Photo: Darryl J. Strickler)

studs (see directions above); then apply the exterior siding over the sheathing and install the fascia and trim boards at the top of the wall.

## INSTALL AIR–DISTRIBUTION AND COOLING VENTS

If you have not already done so, this would be a good time to install new doors, windows, vents, or other air passages between the house and the solarspace and between the solarspace and the outdoors. For low (return) vents from the house, consider using foundation-type vents that are sized to fit concrete block or brick courses (roughly 8 by 16 inches [20.3 by 40.6 cm]). These vents are usually inexpensive and have a back-draft damper or movable louvers. They make ideal low vents for a masonry or masonry veneer wall of the house and can also be used as low (intake) vents between the solarspace and the outdoors.

To install a fan or air duct to deliver heated air to the house, select a location near the top of the wall between the solarspace and the house that will enable you to duct the air to appropriate places. For example, if you want to duct the air to the supply side of a forced-air furnace, you could run the ducts through a utility room or closet located next to the south wall. If you want to deliver the heated air from the solarspace to the north side of the house, you might run a duct between the floor joists of the second floor or attic of the house.

Awning-type windows or wooden basement/utility windows work very well as high vents between the house and the solarspace or as intake or exhaust vents between the solarspace and outdoors. Since you should already have framed in the openings for the exterior (cooling) vents, windows, doors, fans, or vents in the solarspace walls and roof, all you need to do now is install the windows, fans, wind turbines, or whatever. But if you are going to add a new sliding glass door, window, or vents in the south wall of the house, you still have some work ahead of you—so follow the directions below, and get on with it.

**To enlarge an existing window or replace it with a sliding glass door,** first determine whether the wall area to be altered has any plumbing or wiring in it that must be rerouted.

5–28. Foundation-type vents set into block course of recycled concrete block stem wall (Photo: Darryl J. Strickler)

(If plumbing or lighting fixtures or outlets are located on the wall, it is a sure bet that rerouting will be necessary.) Then pry off the trim around the window and remove the existing window and frame and the portions of the interior and exterior wall covering necessary to accommodate the new unit.

1. If the wall is frame construction, replace the header and jambs around the old window with headers and jambs that are the appropriate dimensions for the new unit (see fig. 5–29).
2. If the wall is masonry or masonry veneer, remove the blocks, bricks, or stone from the section of the wall where the new unit is to be placed.
3. Rough in the opening for the new unit. The inside dimensions of the rough opening should allow ¼-inch (0.6-cm) clearance on all sides between the framing and the unit; therefore, the opening should be ½ inch (1.3 cm) wider and higher than the unit itself.
4. Set the new unit in place; be sure it is plumb and level. Secure and fasten it to the framing and spray expanding-type urethane foam in the crack between the unit and the framing. Apply pressure-sensitive aluminized tape over the foam after the foam has cured and been trimmed.
5. Insulate the wall and apply interior and exterior wall surfaces and trim around the new unit.

**To add a new window, door, or sliding glass door** to a blank wall section, first determine how the wall is built and whether it has plumbing or wiring in it. (If you need help with this, check with a remodeling contractor.)

1. If the wall is frame construction, locate the studs inside the wall cavity. (Use a magnet to find the nails, tap on the wall, or look for nail lines in the exterior siding.) Remove the exterior siding or veneer, the interior wall surface, and any insulation from the area where the unit will be installed. Brace the wall temporarily with jacks or sturdy lumber, then cut existing studs and add cripples and headers for the new unit. The rough opening should be ½ inch (1.3 cm) wider and higher than the actual outside dimensions of the unit itself to allow ¼-inch (0.6-cm) clearance on all sides of the unit.

2. If the wall is masonry (brick, concrete block, stone, or adobe), determine where the brick and block courses run and how many blocks or bricks must be removed to accommodate the new unit. Add 4 inches (10.2 cm) to the actual width and height of the outside dimensions of the unit to allow space for a frame made from 2-inch (5-cm) lumber (which is actually 1½ inches [3.8 cm] thick) and ½-inch (1.3-cm) clearance on all sides. For a large window or door, you may need to start at the top of the wall and remove all of the blocks or bricks down to the bottom of the window or door sill. Install angle iron or flat iron above the opening to serve as a header for the new unit before you rebuild the wall above the unit. (Check with a masonry contractor if you need help with this part.) Remove any insulation and interior wall finish from the area where the new unit will be placed.

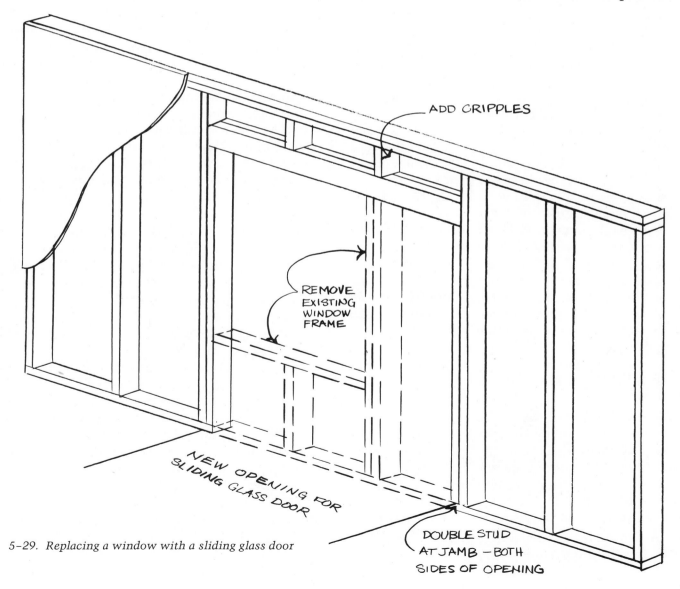

5-29. *Replacing a window with a sliding glass door*

Build a frame for the new unit and attach it to the masonry with lag screws (with washer) screwed into lead shields. Countersink the screw heads and washers. (If there is a frame wall on the inside of the masonry wall, rough in the opening for the new unit as described for frame walls above.)

3. Install the new unit in the rough opening; be sure it is plumb and level. Secure and fasten it to the framing and spray expanding-type urethane foam into the crack between the unit and the framing. Apply pressure-sensitive aluminized tape over the filled crack.

4. Finish the interior and exterior surfaces of the wall around the new unit and apply trim.

**Frame in and install high or low vent openings** through the wall between the house and the solarspace. Follow the instructions above for windows to frame in the vent openings. Consider using standard-size heating and air-conditioning registers with operable registers to cover vent openings. For proper airflow, dampers in high vents must open into the house; dampers in low vents must open into the solarspace. You might also consider constructing high and low vents between the house and solarspace by simply framing in openings cut through the wall and installing hardware cloth, screening material, or a standard register cover and a flap made from 3 mil polyethylene (or tightly woven silk-type fabric). Apply the screen or register cover to the solarspace side of high vents and tape the flap on the inside of the screening so the flap will sail into the house. Apply the screen or register cover to the house side of low vent openings and tape the flap on the solarspace side of the screen so it will sail into the solarspace (see fig. 5–30).

## INSTALL THE GLAZING

Before you install the south wall glazing in the solarspace, clean up the construction debris inside the solarspace and fill the planting beds with soil. (Planting beds may be filled with soil that is one part topsoil, one part sand—or vermiculite or perlite—and one part peat moss. See chapter 6.) It is much easier to shovel soil or

debris through the unglazed south wall than it is to truck it in a wheelborrow after the solarspace is fully enclosed.

After you install the glazing, you will have a fully enclosed functional solarspace, even though it will not be quite finished on the inside. If you have followed the directions up to this point, the solarspace should be ready for the glazing. If the glazing will be installed between the rafters, you will first need to build the interior glazing stops; if it will be installed outside the mullions (or rafters for roof glazing), no interior stops are needed since the mullions or rafters will serve this purpose. (See fig. 5–15 if you need a quick review of glazing details at this point.)

**If you are installing patio door replacement units between the mullions,** build the interior glazing stops on the inside of the mullions and the sill and header plates. Check with the glass supplier beforehand so you know the *exact* thickness and dimensions of the double-insulated units you plan to install; then be sure to allow enough space between the front surface of the interior stop and the outside edges of the mullions, header, and sill plates. Allow an extra $1/16$ inch (0.2 cm) to apply glazier's tape to the stops. Also allow at least $1/4$-inch (0.6-cm) extra space around all four edges of the units to allow for glass expansion and contraction. Remember: glass *moves*—even when it is fixed in place.

1. Apply glazier's tape (Butyl bedding tape) to the edges of the mullion, header, and sill stops on which the glass will rest. Place two or three setting blocks (small neoprene or rubber wedges) on each sill.

2. Set the glass units in place *very* carefully. Make sure they are aligned properly so that at least $1/4$ inch (0.6 cm) of free space remains around all four edges of the glass. (At least two or theee able-bodied people are needed to handle this task gracefully.)

3. Apply glazier's tape around all four edges of the front surface of the glass unit and install exterior stops (trim boards or aluminum division bar) with brass, aluminum, or stainless steel screws. The trim should be screwed in place, since you would not want to take a chance of breaking the glass with

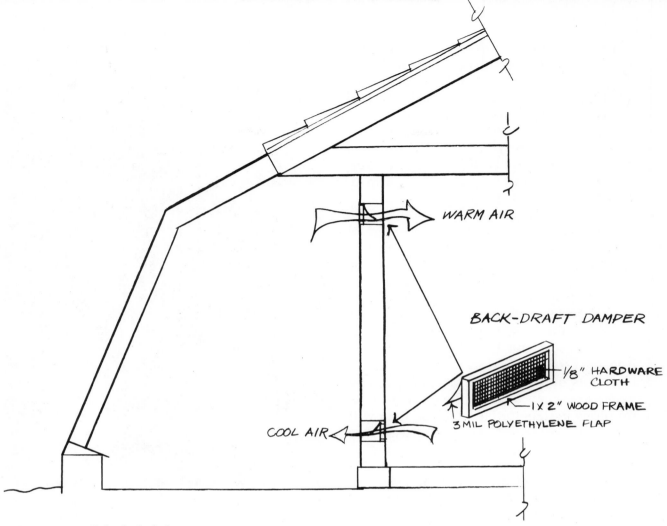

WARM AIR

BACK-DRAFT DAMPER

⅛" HARDWARE CLOTH

1 × 2" WOOD FRAME

3 MIL POLYETHYLENE FLAP

COOL AIR

*5–30.  Vents with back-draft dampers*

an errant hammer blow at this point. (Fastening the exterior stops with screws will also allow you to change the glass should it be broken by an errant hammer blow sometime next year.)

**If you are installing patio door replacement units (or other fixed-size units) outside the mullions,** first install the lower exterior glazing stops. If you use ½-inch (1.3-cm) dowels as the lower stop, drill and glue four dowels for each unit into the sill plate at an angle perpendicular to the glazing surface. If you use 1-inch (2.5-cm) aluminum angle (L shaped), fasten one edge of it to the sill plate with lag screws so that the bottom edge of the glazing unit will rest squarely on the other edge of the aluminum angle. An 8- to 10-inch (20.3- to 25.4-cm) length of aluminum angle, centered under the glazing unit, will usually hold the glazing in place, but for ap-

pearance sake, you may want to run the angle under the entire length of the glazing unit.

1. Apply glazier's tape (Butyl bedding tape) to the outside edges of each mullion, along the bottom of the outside edge of the header, and along the top of the outside edge of the sill plate. Apply a length of ⅜-inch (0.1-cm) square, adhesive-backed foam weather stripping down the center of each mullion. (This is optional; see fig. 5–15.)
2. Set the glass units in place very carefully; be sure they are aligned properly. Equal space should be left between adjoining units along their full length (at least ½ inch [1.3 cm]).
3. Apply glazier's tape along the edges of exterior glass stops and along the top of the front surface of each glass unit. Install the exterior stops (trim boards or aluminum division bar) and fasten them with brass, aluminum, or stainless steel screws.

**If you are installing fiberglass, acrylic, or other roll glazing to the outside of the mullions or rafters,** carefully measure and cut the glazing to the appropriate size so each run of glazing will overlap in the center of a mullion or purlin. Roll glazing, such as Tedlar-coated fiberglass and fiberglass-reinforced acrylic, should be installed during warm weather only. If it is installed during cold weather, it may sag or bow from expansion when the weather turns warm again.

1. Run a bead of caulking or a double width of glazier's tape in the center of the outside edge of each mullion and in the center of the front edge of the header and sill plates. Also run a bead of caulking between the overlapping seam of adjoining runs of glazing (see fig. 5-36). Or, as an alternative to caulking, use Maxi-Seal and Edge Seal gaskets (see Appendix for supplier).

2. Use nails or screws (with a rubber gasket under the head) to fasten the glazing to the mullions, rafters, or purlins if no trim will be installed over the nails. Otherwise, use glavanized nails or staples, as shown in figure 5-36.

3. If two layers of glazing are to be installed, add the second layer to the inside edge of the mullions or rafters in the manner described above (see fig. 5-36). Polyethylene sheeting (with a UV degradation retardant) or a second layer of the same material used as exterior glazing may be installed on the inside of the rafters or mullions as shown in figure 5-36. Woven Poly, a relatively inexpensive triple-laminated polyethylene glazing material, is an excellent light-diffusing inner (or outer) glazing. Woven Poly can be installed very easily with Poly Fastener, a special PVC fastening system that has two interlocking parts (see Appendix for supplier).

# CAULK AND WEATHERIZE THE STRUCTURE

Before you put the finishing touches on your solarspace, go over the entire structure (with a large magnifying glass if necessary) and seal every crack and crevice to keep out air, water, and insects. This should not only include the

5-31. Glazier's tape being applied to outside edges of mullions (Photo: Darryl J. Strickler)

5-32. Glazier's tape in place on outside of mullions; glass will rest on setting blocks flush with outside wall of solarspace, on top of custom-made flashing (Photo: Darryl J. Strickler)

5-33. Patio door replacement units being installed as solarspace glazing (Photo: Darryl J. Strickler)

5-34. Glazier's tape is applied to underside of exterior trim boards (Photo: Darryl J. Strickler)

5-35. Exterior trim is applied between glazing units (Photo: Darryl J. Strickler)

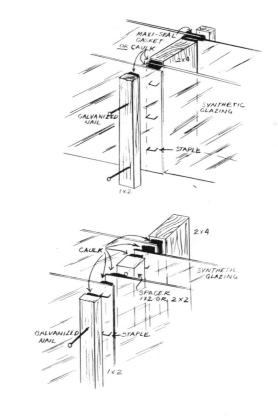

5-36. Mullion detail for installing roll glazing

more obvious things like applying waterproofing sealer around the flashing and roof vents and caulking around the glazing; it should also include spraying expanding-type urethane foam into all the cracks around windows, doors, vents, drainpipes, under the bottom edge of the siding, along the joints where the roof and walls meet, where the solarspace is attached to the house—anywhere you can find a crack or joint large enough to spray into or caulk.

**Apply waterproof roofing compound** along the upper edge of the flashing between the solarspace roof and the house and along the flashing for vents, windows, or skylights in the roof of the solarspace. Also apply roofing compound or caulking along the edge of the vertical flashing between the end walls of the solarspace and the wall of the house. (Roofing compound—a black sticky substance—that comes in caulking-type tubes is easiest to apply.)

**Caulk around the glazing** with a good-quality silicone caulking compound, in your favorite color—usually bronze, brown, black, white, and sometimes tan. Silicone is more expensive but works best for joints that are subject to expansion or other movement. Most types of Butyl will eventually harden and crack and therefore must be replaced periodically if exposed to direct sunlight and continual expansion and contraction. (Some types of Butyl and silicone are reported to interact unfavorably with one another when used in combination, so it makes sense to use Butyl caulk if it comes in direct contact with Butyl tape or the Butyl seals of double glazing units.)

Depending on the specific glazing details used, caulking around glazing may not be necessary. For example, if the glazier's tape under the exterior glass stops appears to have adequately sealed the glass, it may not be necessary to caulk along the edges of the glass. Use your judgment.

1. Run a ¼-inch (0.6-cm) bead of caulking along the edges of the exterior trim boards (glass stops or glazing bars) where they meet the glazing surface *if* you did not use glazier's tape under the trim boards. (The most effective way to apply caulking is to push the caulking gun using steady pressure on the trigger.)
2. Caulk around the edges of window and door trim to keep water, air, and insects from getting behind the trim.

Spray expanding-type urethane foam in all cracks and crevices in the inside and outside of the structure. Spray this foam anywhere you find a crack, but keep it off your hands—it is very sticky and difficult to remove. Do not try to smooth out the excess foam that will ooze out of the crack; wait until it dries, then cut off the ex-

5-37. *Apply roofing compound around flashing and roof openings (Photo: Darryl J. Strickler)*

5-38. *Caulk around exterior trim to keep out bugs and water (Photo: Darryl J. Strickler)*

5-39. *Fill all cracks with expanding-type foam; let it dry, then trim off excess (Photo: Darryl J. Strickler)*

cess with a knife. If the foamed crack will be covered with trim or finish material, cover the foamed-in crack with adhesive-backed aluminized tape.

**Install weather stripping** in window and door jambs. V-type plastic weather stripping with one adhesive edge works very well and is easy to install; follow package directions carefully.

## FINISH OUT THE INTERIOR

How you finish out the interior of your solarspace is largely a matter of personal taste, finances, and function. For example, the interior could be very Spartan, with earthen, gravel, or stained concrete floors and painted walls and ceilings; or it could have an elegant quarry tile or stone floor laid over a concrete slab, wood trim, tongue-and-groove wood ceiling, and stucco walls.

Use the guidelines and suggestions below to create the kind of interior you want (or can afford).

### Floors

A solarspace floor may be finished in many ways, all the way from an *un*floor (your basic dirt floor), to brick over sand, to expensive floor tiles laid on a concrete slab. The options of using a pebble, brushed, or patterned concrete slab floor must obviously be selected *before* the slab is poured. Patterned floors require special equipment (like big cookie cutters) to stamp brick, tile, or stone patterns into the molten concrete before it has dried. The individual "tiles," "bricks," or "stones" in the patterned floor can then be stained a dark color and smoothed; the grout grooves may be left unstained and rough to resemble real grout. You can also cut a tile pattern in a concrete floor with a masonry saw after the concrete has cured. This is somewhat difficult because it requires perfectly straight cuts in two directions to simulate the grout grooves that would be in a 12- by 12-inch (30.5- by 30.5-cm) tile floor, for example. It can, however,

5-40. *Patterned concrete floor, doing a great job of looking like the real thing (Photo: Darryl J. Strickler)*

5-41. Saw-cut concrete floor in Hard/Stevens solarspace (Photo: Darryl J. Strickler)

result in a very interesting floor pattern—especially if the saw wanders off course.

A smooth floor slab can be painted or stained with products specially formulated for concrete floors and/or the floor can be treated with a clear sealer that will darken the surface color somewhat. It is best to wait until the floor has cured ninety days before staining it and to etch the floor with an acid etching solution before painting or staining it. If you want the floor slab to store heat, it should be a dark color, such as a dark, reddish brown (cordovan). The reddish tint will bounce some of the light from the red end of the light spectrum back to the plants in the solarspace. If the wall between the solarspace and the house is an uninsulated masonry wall, you may want the floor in the solarspace to be a lighter color so that it will bounce more light and heat to the house wall. This will allow the wall to store more heat than the floor—theoretically, at least.

Quarry tile, stone, or brick can be laid directly on a concrete floor slab if you are seeking a more expensive look. (It not only looks expensive, it usually is; although if you shop or scrounge around, you may find some excellent buys.) Brick, tile, or stone floors should be a medium or dark color and be left unglazed. Such

floors should be sealed with a specially formulated finish designed for masonry floors that will produce a satin rather than a high-gloss finish. Slate can also be used if you already have some or can find it very cheap, but it tends to overheat when struck by direct sunlight and give off its stored heat more quickly than brick, stone, or quarry tile.

A relatively low-cost alternative to brick, tile, or stone that will produce an interesting-looking floor and add thermal mass at the same time is 2- by 8- by 16-inch (5- by 20.3- by 40.6-cm) solid concrete cap blocks laid in any pattern you desire—two by two, herringbone, or whatever. Cap blocks are laid without mortar joints, either directly on the floor slab if it is very level or over a layer of thinset concrete. The entire surface of the blocks is then washed with thinset or a grout mixture to smooth out the porous texture of the blocks. You can either use a premixed color in the grout or thinset or stain the surface the desired color.

## Walls

Insulated end walls in the solarspace should have a light-colored, or reflective, surface; green or red walls also work well. Interior masonry walls struck by direct or reflected sunlight should be a dark color to increase absorption of light and heat. Masonry walls in indirect sunlight, such as the inside surface of the stem wall, can be any color. Light-colored wood, such as pine or Alaskan yellow cedar (alderwood), with a smooth surface and a glossy urethane finish effectively reflect light if you prefer (and can afford) to have wood instead of Sheetrock (drywall) on the side walls, for example. Of course, Sheetrock is a cheaper solution for covering walls, and it is easy to install and paint. It is *not*, however, easy (or pleasant) to seal and sand around the joints. To skip the tedious work (and choking dust) of sanding the joints, nail the Sheetrock in place, then tape the joints and trowel Sheetrock joint compound over the entire surface of the Sheetrock panel to produce a stucco-type finish. (Use only ⅝-inch (1.6-cm) water-resistant Sheetrock to help prevent sagging that can result from high moisture conditions.) Use more than one layer of Sheetrock if you want to add interior mass to the solarspace—that is, nail one layer on top of another.

Redwood, cedar, and cypress are moisture-tolerant woods and are therefore ideally suited for covering walls and for use as trim in a solarspace. They may not be ideally suited to your finances, however. These woods are relatively expensive, and they do not reflect light very well. In a large solarspace, the reflectance is not as important as it is in a smaller space where you would want to reflect as much light as possible to plants and thermal storage mass.

If wood is used to cover end walls and as trim, use the same variety of wood throughout the solarspace. For example, beveled redwood siding could be installed across the end walls on an angle to match a sloped glazing surface and could also be used as interior glazing stops. In addition, redwood lumber could be used for plant shelves or decking around a soaking tub. This one wood approach gives the interior of the space a more unified appearance. (Do not paint redwood, cedar, or cypress. Their natural appeal needs no additional help.)

## Mullions and Rafters

The finish treatment for exposed mullions and rafters depends somewhat on whether the glazing has been installed between or outside of them. If the glazing is set between the mullions and rafters, interior glazing stops are required. These can either be constructed of standard finishing lumber, such as pine or fir that has been stained or painted, or they can be built from more expensive woods. If the glazing is set on the outside of mullions and rafters, you need only paint or stain the rafters or mullions, but you could cover them with wood trim to hide the imperfections in the framing lumber. Another (expensive) possibility—one that would have to be selected early in the construction process—would be to use clear (no knots) fir or redwood (select structural grade) for exposed mullions and rafters. This wood needs only a clear sealer to present a finished appearance. The added cost of select structural–grade lumber used as framing may be worthwhile if it eliminates the need for expensive wood trim over the exposed mullions and rafters.

## Ceilings

The ceiling of a solarspace should be a smooth, light-reflecting surface if you are planning to grow food or plants in the space. Water-resistant 5/8-inch (1.6-cm) Sheetrock painted white works well as a ceiling covering, but light-colored tongue-and-groove wood, such as "car siding," with a semigloss finish or even rough cedar exterior siding may be used to add a little class. Another possibility that your plants, if not your Aunt Harriet, will probably be wild about, would be to nail Sheetrock, or any water-resistant material with a smooth surface, to the ceiling, but instead of finishing it in the usual way, cover the ceiling surface with an adhesive-backed aluminized film. Not only will this film reflect light to the plants, but it will also serve as a radiant barrier to keep the heat in the solarspace and prevent moisture from penetrating the insulation in the ceiling. If you can appreciate a utilitarian look, forget the Sheetrock and simply nail foil-faced insulation or sheathing board under the rafters and tape the joints with aluminized tape. Another alternative would be to use a thin layer of translucent plastic material such as Plexiglass as an inner roof glazing (under the rafters) and continue it under the insulated portion of the roof. This would give a very unified (albeit very plastic) appearance to the ceiling of a solarspace. The possibilities for finishing the ceiling—or the remainder of a solarspace for that matter—are limited only by your imagination and your budget.

## . . . THERE YOU HAVE IT

Congratulations—you have now completed your very own solarspace. Your good judgment, sound planning, excellent design work, and fine craftsmanship have paid off and are to be much admired and appreciated. So stand back and admire it; take some photographs to send to the folks back home; invite your friends and neighbors (who have been staying away for fear you would put them to work) to come over for the christening party. But . . .

But what then—what are you really going to *do* with it? Now what?

# SIX

# Now What?

*Now that you have completed your solarspace, what are you actually going to do with it, or in it?*

How you use your solarspace will significantly influence both your satisfaction with it and the amount of time it takes to recover your investment. Your investment can be recovered not only in dollars and cents saved on heat and food, but in enjoyment of life-enhancing activities. If you designed and built your solarspace for specific purposes and you actually use it for those purposes, you should have no difficulty reaping the benefits your solarspace has to offer. On the other hand, if you have not yet devoted too much serious thought to how you will use, operate, and maintain your solarspace after it is completed, you should do so now. So read on. In this chapter you will learn how to make the most of your solarspace.

## GROWING INTO YOUR SOLARSPACE: OPERATION AND MANAGEMENT

If your solarspace was designed primarily for heat production, you will operate it differently than you would if you had planned it primarily for food production, for example. If your major interest is to use your solarspace as added space for living-related functions, the most important consideration will be to keep it at a comfortable temperature year-round or during the times you want to use it.

As you will discover, the three functions of a solarspace—heat production, food production, and added space—can run at cross purposes to one another when you try to achieve them all at the same time. Through careful management and operation, however, your solarspace should perform at an optimum level for its primary function and, with some compromise on your part, should also adequately serve whatever secondary purposes you may have in mind.

## Regulating and Distributing Heat in Winter

If you are going to use your solarspace for growing plants, for living-related functions during the heating season, or for both, the total amount of heat produced in the solarspace on sunny days must be divided between the heating requirements of the house and the needs of people and plants inhabiting the solarspace. The amount of thermal mass and the color and absorption rate of interior surfaces and plants in the solarspace will determine, to a great extent, how much heat will be available to supplement the home-heating demand. For example, the leaves of plants in a solarspace will absorb a considerable amount of the available sunlight, which is essential for their growth during the winter when the hours of available sunlight are few. Plants must, therefore, compete with thermal mass for the available sunlight. The thermal mass must, in turn, receive an adequate amount of direct sunlight so it can reradiate its stored energy to keep the plants (and people) comfortable when the sun goes down.

During a sunny day in winter when the sun produces more heat in a solarspace than can be absorbed by plants and thermal mass—and at a rate faster than the absorption rate—the air temperature in the space will rise. If the heated air is not drawn off by natural convection or by fans for use in the house and the air in the solarspace becomes overheated, the excessive heat will eventually be absorbed by plants, soil, and thermal mass, or it will be lost to the outdoors through the glazed areas and the insulated roof and walls of the solarspace.

What all of this means in a practical sense is that you need to open doors, windows, and vents in the wall between the solarspace and the house when the air in the solarspace is warmer than the air in the house. When the air in the solarspace is cooler than the air in the house, you simply close them again. Two maximum-minimum thermometers—one in the solarspace and one in the house—will help you decide when to open and close doors, windows, and vents. (If the vents have back-draft dampers as described in chapter 5, they will work automatically.) To save steps comparing thermometer readings between the house and the solarspace, you might install an indoor-outdoor thermometer inside the house, with the outdoor bulb and conduction wire running into the solarspace. Or, to entirely eliminate thermometer reading, you could install automatic vent operators, which will open and close vents, or thermostats to control fans. (The Appendix lists several manufacturers of automatic control devices.)

Any of the air-distribution systems described above will work satisfactorily. When there is no heat to distribute, however, such as during an extended period of cold, cloudy weather or during subzero weather when the solarspace cannot produce and retain heat as rapidly as it is being lost, you may have to supply supplementary heat to the solarspace if you want to keep plants—and people—comfortable. A small wood-burning stove or portable, electric, fan-forced heater could be used during the most extreme weather conditions. A door to the house could also be left ajar so the solarspace could "steal" heat from the house. This latter approach should be used only if the solarspace has movable insulation over the glazing—otherwise the heat from the house will be readily lost to the outdoors via the solarspace. A fan-control system such as the Wesper 2 pictured in figure 6-1 can be set to extract warm air from a solarspace during the day *and* return warm air to the solarspace at night.

If the solarspace has no added thermal mass (beyond that contained in is framing, glazing, and wall-covering material) and you do not plan to inhabit it regularly or grow plants or food in it during the heating season, all the heat it pro-

6-1. *Weather Energy System's Wesper 2 fan and control system (Photo: Darryl J. Strickler)*

duces can be used to heat your house. In this case a fan and duct system should be used to distribute solar-heated air to the house or a rock storage bed as previously discussed. The simplest arrangement is to install an in-line thermostat high on the north wall of the solarspace so that it will activate (and deactivate) the fan(s) at a predetermined temperature—say 75°F (23.9°C)—to distribute heat to the house. (This thermostat should be disconnected during the cooling season if it does not have a set-point high enough to guarantee that the fans will not be activated.) A system more elaborate than the one described above can, of course, be devised or purchased if desired.

## Reducing Heat Loss through Glazing in Winter

Many owners and designers of solarspaces regard the development of methods and materials to reduce heat loss through glazing areas as a challenge to their creative abilities. In fact, entire books have been devoted to this topic alone (see Appendix), and new schemes, materials, and commercial products for insulating glazing are continually being developed. Commercially available products include multilayer aluminized-film curtains, quilted insulating curtains, and movable insulated louvers. Because most of these products are rather expensive, many solarspace owners find it necessary to invent their own means of reducing heat loss through their solarspace glazing. Financial considerations notwithstanding, such invention is often the daughter of necessity because no two solarspaces or owners are exactly alike. Therefore, an individualized approach must be developed for reducing heat loss through the glazing area of a specific solarspace—in this case, yours.

Although you can certainly plan ways to reduce heat loss through glazing areas *before* you build your solarspace, you may discover after you have used your solarspace for several cold months (or after you have run out of money) that you can manage satisfactorily without insulation for the glazing areas. You may also discover, as many other people have, that what you planned will not work because of some structural feature or because of the arrangement of plants and furnishings—or some other unforeseen reason. Therefore, it is often necessary to actually build and inhabit your solarspace before you can make a truly informed decision on the type of movable insulation (if any) you need for your solarspace.

As is the case with so many things in life, you have two basic choices: you can either try to slow down the heat loss through your solarspace glazing at night and during overcast days in winter, or you can let the heat escape. Whether you will absolutely need a means to reduce heat loss through the glazing depends on how and when you use your solarspace; on the number of layers of glazing it has; and on the amount of winter sun and number of heating degree days in your area.

### Climate

In geographic regions that have 60 percent (or more) of possible sunshine and relatively mild winters (under 4,000 degree days), a solarspace with double glazing and adequate thermal mass can be expected to remain within an acceptable temperature range for plants and people—except during the most severe weather conditions—without glazing insulation. In such climates it is usually more cost-effective to run an auxiliary heater—say a thirty-dollar portable electric heater—in the solarspace on the few nights a year it would be required than it would be to purchase insulating curtains, for example.

Although there is not much direct solar radiation during the winter months in extremely cloudy climates like the Pacific Northwest, the cloud cover tends to moderate the climate and reduce radiant losses from the earth (and solarspaces) at night. This, in turn, helps reduce the need to insulate glazing at night in solarspaces built in such climates. Solarspaces located in areas of abundant winter sunshine, such as parts of Oklahoma, New Mexico, and Colorado, often gain enough heat during sunny days to offset nighttime losses, despite some very cold temperatures. Therefore, it is often possible to maintain habitable temperatures in solarspaces in such climates without insulating the glazing at night if the solarspace has adequate thermal mass. In areas with cold, cloudy winters, such as the upper Midwest, the North Atlantic states, and New England, movable insulation for solarspace glazing is usually a necessity if the space is to be inhabited after sundown by people or plants during the winter months. If a solarspace located in these areas—or anywhere else for that matter—is used only for daytime heat production during the winter, however, movable insulation is usually unnecessary. A low-mass solarspace can fluctuate greatly in temperature during the winter months if no plants are in the space. A solarspace with plants (or people) in it should remain within the range of 50° to 85°F (10° to 29°C) throughout the year.

## Cost and Practicality

No matter where you live, insulating or radiant devices placed over the glazing of your solarspace (or house) will greatly reduce heat losses at night and during overcast days in winter. So why don't all owners of solarspaces use glazing insulation? As previously noted, it is expensive: two to five dollars a square foot if you make it yourself; four to twelve dollars per square foot—plus the cost of installation—if you purchase it ready-made. Although you can cer-

tainly make glazing insulation that is both attractive and efficient, it is very difficult to duplicate the appearance and thermal efficiency of commercially produced insulating systems. If you can afford ready-made insulating systems, the expense is not difficult to justify (it also qualifies for tax credits); if you cannot, you will want to investigate making your own. Some types of homemade (and commercial) glazing insulation can end up looking rather unsightly and cumbersome when they are in place. They may also present problems with installation, especially on sloped glazing surfaces, and require time (and patience) to put in place and remove, or raise and lower. Despite these possible disadvantages, you may find that some type of insulating or radiant device on glazing surfaces is necessary if you want to keep your solarspace comfortable year-round.

Literally hundreds of schemes, materials, and manufactured products have been designed to reduce heat loss through glazing. The cost and practicality of these items varies greatly. Obviously, they cannot all be described here. The recommendations for constructing and instal-

6–2. *Insulating Curtain Wall (ICW) (Photo: courtesy of Thermal Technology Corporation, Broomfield, CO)*

ling insulating or radiant devices or both presented below are intended as a starting point to provide some suggestions for what have proved to be the most practical and cost-effective methods to reduce heat loss through the glazing areas of solarspaces. For example, you will not find any information on exterior insulating devices because such devices are costly, difficult for most people to construct, and subject to wind and snow damage. (You can find information on external shutter systems for solarspaces in the sources listed in the Appendix, along with a great deal of other helpful ideas for glazing insulation not presented below.) I leave you to your own devices, because you will have to make your own decisions based on your specific requirements, tastes, and financial situation.

### Press-fit Rigid Insulation

Panels of foil-faced (or nonfaced) rigid-board insulation can be cut to fit tightly over glazing surfaces (such as between mullions or rafters) and held in place with adhesive-backed magnet tape or Velcro strips. For press-fit insulating panels, Styrofoam (Blue Dow), urethane, or isocynurate rigid insulation, such as Thermax or High-R, are preferable to white polystyrene beadboard. Beadboard tends to crumble and deteriorate around the edges from repeated handling and press-fitting, although tape or aluminum extrusions around the edges help to reduce this problem.

Insulation board can be applied with construction adhesive to a rigid material such as Masonite or heavy, gray industrial-grade cardboard to increase its strength and durability. It can also be painted with latex-base paint, covered with a decorative fabric, designer bedsheets, wallpaper, contact paper, or any other material of your choice. How about this: glue the insulation board to a sheet of Masonite, then cover the Masonite with a large wallpaper-type mural or paint it to match the interior of the solarspace.

One of the more obvious disadvantages of using rigid panels over glazing areas is the problem of storage when the panels are not in use. This can be solved by building a rack under the insulated portion of the roof of your solarspace or against the east or west walls to store the insulation when it is not on the glazing. (Storing

the panels in this manner will further increase the R value of the roof or walls.) Another problem neither so obvious nor so easily solved is that some types of rigid insulation board produce toxic fumes if they are ignited. That is not so likely to happen unless your whole house catches fire and, of course, other common household items also give off toxic fumes. You should, however, be aware if your insulation *can* give off toxic vapors; before you purchase a given type of rigid insulation board to use for glazing insulation, check on its combustion properties. (Some manufacturers tend to conceal this kind of information in a blizzard of technical jargon and formulas.) Klegcell insulation board has a high R value per inch; it holds up well under repeated handling and does not produce toxic vapors when ignited.

### Roll-down Insulating Curtains

Various types of roll-down insulating or radiant/reflective curtains can be purchased from the manufacturers and suppliers listed in the Appendix. These include multilayer aluminized curtains, such as the ICW (fig. 6–2), which have air spaces between the layers to increase the R value of the curtain; and fabric-covered quilt-type insulating curtains or Roman shades, such as Warm Window (fig. 6–3).

If you prefer to make your own insulating curtains, follow the plan for making the PSDS insulating curtain illustrated in figure 6–4. Start with a layer of decorative fabric for the inside; designer bed sheets work well for this purpose, but any lightweight, tightly woven fabric will do the job. Next, include a layer of Astrolon, Mylar,

*6-3. Warm Window insulated Roman Shades (Photo: courtesy of Warm Window, Seattle, WA)*

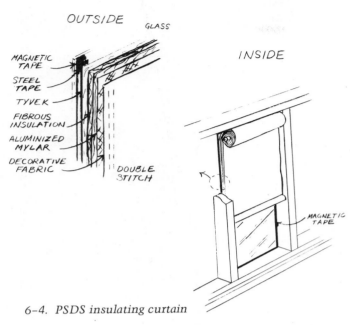

PSDS INSULATING CURTAIN

OUTSIDE

GLASS

INSIDE

MAGNETIC TAPE

STEEL TAPE

TYVEK

FIBROUS INSULATION

ALUMINIZED MYLAR

DECORATIVE FABRIC

DOUBLE STITCH

MAGNETIC TAPE

6-4. PSDS insulating curtain

polyethylene, or other vacuum-deposit aluminized film to serve as a radiant/vapor barrier; then include a layer of Thinsulate, Hollofil, Polarguard, or other fibrous insulating material designed for clothing and bedspreads. A mattress pad or a sheet of white packing foam may be used in place of the fibrous insulation. Finally, for the exterior layer, use Tyvek, a spunbound polyolefin that resembles tough, clothlike, white paper. Tyvek is an excellent air-infiltration barrier, but because it "breathes," it will not allow moisture to condense in the curtain insulation. Tyvek should, however, be kept out of direct sunlight, since ultraviolet radiation will eventually cause it to deteriorate. It is no accident that the design of the PSDS insulating curtain described above simulates the construction materials used in the walls of your solarspace. The curtain will work well to hold heat in your solarspace for the same reasons your walls will.

Materials for the PSDS curtain are readily available at a moderate price. Blanket-size aluminized film can be purchased in most backpacking or outdoor stores under such names as emergency blankets, space blankets, or thermos blankets. Larger-size sheets of Astrolon, aluminized polyethylene, and Mylar may be purchased by mail from Shelter Institute

(see Appendix). Most larger fabric stores sell fiber-fill insulating or quilting material, or they can order it for you. You can buy mattress pads at your local discount department store, and white packing foam is available on order from many of the larger office supply stores. Tyvek can be ordered from Du Pont or PARSEC (see Appendix) under the trade name Airtight-White or purchased from a building materials supplier.

A less expensive version of the PSDS insulating curtain can be made by substituting a layer of bubble pack for the layer of fiber fill. Bubble pack is a low-cost polyethylene packing material with air bubbles sealed into it (that kids of all ages love to pop). It can be used by itself, or with a layer of aluminized film, to make a very functional and inexpensive (if not beautiful) insulating curtain. Products like Air Cap SC 120 are available from commercial suppliers in large sheets (see Appendix).

A layer of bubble pack and a layer of aluminized film taped together around the edges to produce a finished seam, although allowing some dead air space between the film and the bubble pack, can result in an excellent radiant/insulating curtain. A few holes should be punched along the bottom edge of the aluminized film to allow air to escape when the curtain is rolled up.

Some manufacturers make adhesive-backed bubble pack—Vikalite or Thermolite—for use as inexpensive, stick-on storm windows. By attaching a layer of aluminized film directly to the adhesive backing of this material, you can make an insulating curtain very quickly. Aluminized film is also manufactured with adhesive backing (PARSEC "Thermo-Brite") so you can apply regular bubble pack to the aluminized film if you prefer. You can, of course, also apply nonadhesive aluminized film to nonadhesive bubble pack with a product like 3M Scotch Grip Plastic Adhesive 4693. Foil-Ray is a commercially available product that is composed of a layer of bubble pack already laminated to aluminum foil.

Bubble pack can also be used alone as insulating curtain. Although it does not have a significant R value, bubble pack is useful on cloudy days and even in summer because it provides some insulating qualities while admitting light for plants. It also affords some privacy since it refracts light. A roll-down curtain made

6-5. *Bubble pack insulating roll-up curtain (Photo: Darryl J. Strickler)*

from bubble pack, with a rope-and-pulley system and a weighted rod such as that used on porch curtains, is rather easy to make.

### Edge Seals

No matter what type of insulating or radiant curtain or roll-down shade you buy or make to reduce heat loss from your solarspace glazing, the edges of the curtain or shade should be sealed against the glazing framework or to the glazing itself to prevent airflow between the curtain or shade and the glazing surface. Under certain conditions an insulating or radiant curtain or shade that allows air to flow behind it can actually *increase* rather than decrease heat loss through the glazing. A simple solution for sealing the edges of curtains or shades (and for attaching rigid, press-fit insulation to glazing surfaces) is to use ½-inch-wide (1.3-cm) adhesive-backed magnet tape such as 3M Plastiform Magnetic Tape. Velcro also works well for edge seals but is slightly more expensive. An airtight seal can be achieved by attaching one strip of magnet tape or Velcro to the edges of the glazing framework (or the glazing itself) and another strip along the edge of the curtain or shade. Adhesive-backed steel tape (available from Shelter Institute) works better than magnet tape on the edge of a roll-down shade because it is not as thick and will roll up more easily. (It is also less expensive.) The strips of tape or Velcro do not need to be continuous if you are trying to cut costs, but a full run will produce a better seal. Another solution to reduce the amount of magnet tape needed is to attach a thin strip of ferrous metal to the glazing framework and use the magnet tape only on the edges of the curtain

6-6. *Magnetic tape edge seals on roll-up radiant shades (Saccone residence; photo: Darryl J. Strickler)*

or shade. Or use adhesive-backed steel tape on the edges of the shade or curtain and magnet tape on the glazing framework. Wind-N-Sun Shield, Inc. makes an edge-seal system called Wind-Stop, which is ideally suited for roller shades (see Appendix).

### Radiant Curtains and Shades

Aluminized films, such as Astrolon I or VIII, or aluminized fabrics like Foylon 7194 or 7041 provide an excellent barrier against radiant heat loss through solarspace glazing. Aluminized films and fabrics and spunbonded polyolefin materials like Tyvek have been shown to reduce the cost of heating large commercial greenhouses by as much as 57 percent. Because these materials are relatively inexpensive, lightweight, and highly compressible, they are ideally suited for making radiant curtains for a solarspace. Such curtains can greatly reduce excessive radiant heat loss from a solarspace, particularly on clear cold nights, that results from night-sky radiation.

Radiant curtains can be made from two or more layers of aluminized film such as Astrolon; or two layers of Tyvek with an intermediate layer of Astrolon; or one or two layers of polyester-backed aluminized fabrics like Foylon

7194 or aluminized Mylar film; or a single layer of foil-faced fabric like Foylon 7438 or Archimedes Shield, sold by Suncraft (see Appendix). (FOYLON fabrics have one bright and one dull side. When using multiple layers of this material, the bright side of each successive layer should face in the same direction; air spaces between successive layers should be bounded by one dull surface and one bright surface.)

Although figure 6–7 illustrates one method of installing a radiant curtain in a solarspace with sloped south wall and roof glazing, this method will also work for solarspaces with other glazing configurations. It could also be adapted for heavier insulating curtains. Figure 6–8 shows a versatile scheme for installing radiant curtains in a solarspace with no roof glazing. Figure 6–9 illustrates a radiant shade attached to a conventional pull-down blind roller. This shade is made from a standard window shade with a layer of aluminized film attached to the shade with an adhesive like 3M Scotch Grip Plastic Adhesive 4693. Materials like Dura

Shade 4413, Foylon 7192 or 7082, PARSEC Vapo-Brite (aluminized film laminated to polyolefin) or Archimedes Shield work very well for making your own pull-down shades on spring-loaded rollers. The type of shade illustrated in figure 6–9 is particularly useful for east and west wall glazing and for doors and windows, but it can also be used on a southern wall or roof glazing. You may want to design your radiant shades to be reversible. This will allow you to face the aluminized side in during the winter and out during the summer.

### Managing Glazing Insulation

Insulating or radiant curtains and shades, and press-fit rigid insulation, are placed (or rolled down over) glazing surfaces at night and during extremely cloudy or stormy days in winter, and are removed (or raised) when the sun is visible during the heating season. Motorized controls and light- or heat-sensing devices may be installed to operate glazing insulation if desired.

6–7. *Side-pull PSDS radiant curtain for solarspace with south wall and roof glazing*

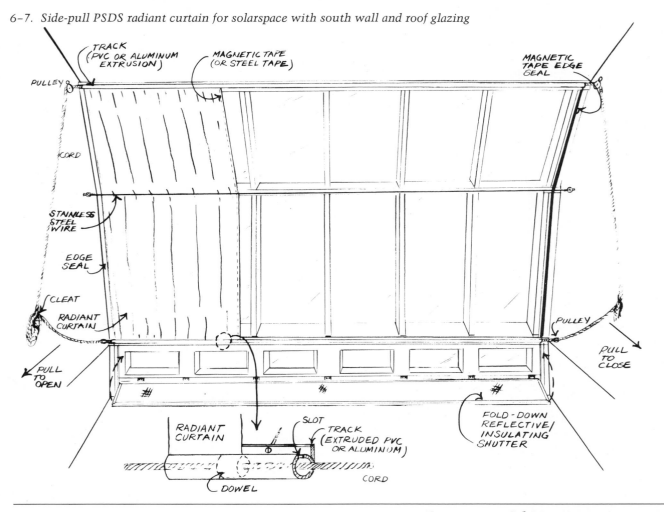

**6-8.** *Drop-down radiant curtain/roof reflector for solarspace with no roof glazing*

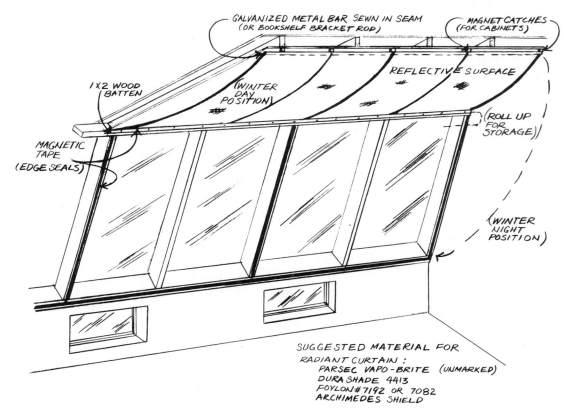

GALVANIZED METAL BAR SEWN IN SEAM
(OR BOOKSHELF BRACKET ROD)

MAGNET CATCHES
(FOR CABINETS)

REFLECTIVE SURFACE

1X2 WOOD BATTEN

(WINTER DAY POSITION)

(ROLL UP FOR STORAGE))

MAGNETIC TAPE (EDGE SEALS)

(WINTER NIGHT POSITION)

SUGGESTED MATERIAL FOR
RADIANT CURTAIN :
PARSEC VAPO-BRITE (UNMARKED)
DURA SHADE 4413
FOYLON #7192 OR 7082
ARCHIMEDES SHIELD

**6-9.** *Pull-down radiant shade*

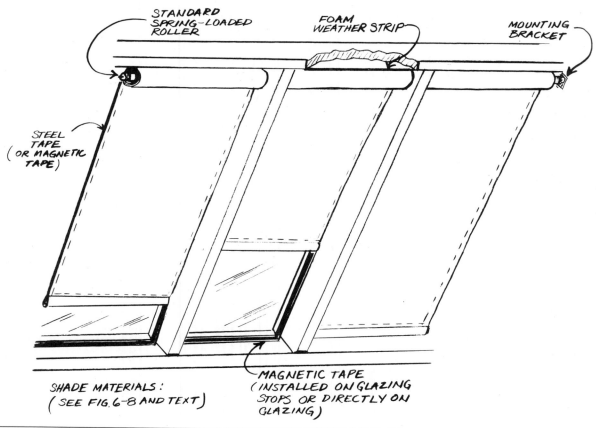

STANDARD SPRING-LOADED ROLLER

FOAM WEATHER STRIP

MOUNTING BRACKET

STEEL TAPE (OR MAGNETIC TAPE)

SHADE MATERIALS :
(SEE FIG. 6-8 AND TEXT)

MAGNETIC TAPE
(INSTALLED ON GLAZING STOPS OR DIRECTLY ON GLAZING)

## Reducing Heat Gain through Glazing in Summer

Direct sun in the interior of a solarspace during the cooling season can result in excessive temperatures, making the solarspace a downright uncomfortable place for people and plants. An adequately sized ventilation system, as described in chapter 3, will exhaust hot air and help reduce the interior temperature of a solarspace, but your first line of defense is to block the sun from entering the space. It is especially important to keep your solarspace cool if you normally use fans or air-conditioning to cool your house. An overheated solarspace attached to your house can potentially increase your cooling costs enough to cancel the energy savings from the heat your solarspace produces in winter—or even enough to increase your total energy use over the year—because of the increased cooling load. On the other hand, a solarspace can help reduce your cooling bill if the space is cool and shaded, since it serves as a buffer between the interior of the house and the hot summer sun in the southern sky. So keep the sun out of your solarspace in summer and open the cooling vents. Experiment a bit; you may find that it is best to close off the solarspace from the house in summer, leaving only the cooling vents open.

Try out several strategies to find the best approach to shading your solarspace. If deciduous trees grow to the south, east, or west, see how much natural shade they provide and monitor the interior temperature of your solarspace for several weeks. If you incorporated a properly sized roof overhang on your solarspace and are relying upon it to provide shading, see how well it works; see if the heat gain from the low morning and late afternoon sun is tolerable. If you are planning to use natural shading from vines growing from the ground or on a trellis, be sure you plant the vines early enough in the season (after the last frost) so they will provide enough coverage to shade the glazing of your solarspace when you need shade. Should you discover that these natural means or structural features of your solarspace do not block the sun adequately, start shopping for external shading devices.

As previously discussed, greenhouse shade cloth—variously called sunshade screening, solar screen, and so forth—is the best solution for external shading applications. Shade cloth is a woven screening material, usually fiberglass or polypropylene, that comes in a variety of widths and colors, including decorative stripes, and in a range of light-blocking capabilities (see Appendix for manufacturers). Depending on the particular type of shade cloth you select and the quantity your purchase, shade cloth costs between thirty-five cents and one dollar per square foot—which makes it a very good buy in relation to the amount of heat it blocks from entering your solarspace. Shade cloth that blocks 70 to 80 percent of the available sunlight will still admit enough daylight to adequately illuminate the interior of a solarspace, and it does not significantly obstruct the view to the outdoors. If you have any serious crops growing in your solarspace during the cooling season, however, the shade cloth you select should block no more than 25 percent of the sunlight.

You can install shade cloth in many ways. It can be casually draped over the solarspace glazing, made into a roll-down curtain employing a rope-and-pulley system and a weighted rod, or tightly stretched over the glazing and held in place with a clamping system. VIMCO Solar Shield has a simple mounting system that allows the user to roll up the shade cloth during the heating season without removing it from the glazing framework. A roll-up system allows greater flexiblity, which is particularly useful during the spring and fall when periods of warm

*6-10. Shade cloth in place on Lerner solarspace (Photo: Darryl J. Strickler)*

and cold weather alternate. Phifer SunScreen has a channel locking frame system for its shade cloth (see fig. 6-11).

Shade cloth can be easily cut and sewn, and grommets can be installed on lapped seams and corners to anchor the shade cloth to the glazing framework with small hooks or fasten it to anchors in the ground with elastic shock cords. Each glazing unit can be fitted individually, or a large drape of shade cloth can be made to cover an entire glazing area by sewing several widths together. Wider cloth obviously reduces the number of seams necessary to make a large drape. Phifer SunScreen is manufactured in a wide array of colors in widths up to 84 inches (213.4 cm). With the variety of shade cloth available and the number of possible methods of installing it, you should have no difficulty deciding how to use this functional and decorative product.

6-11. *Phifer SunScreen in channel-lock frame system on Liveoak solarspace (Photo, courtesy of Phifer Wire Products, Tuscaloosa, AL)*

One word of caution is in order, however. *Do not* allow shade cloth to come in constant, direct contact with synthetic glazing. Some types of shade cloth can bond permanently to the surface of some types of synthetic glazing. Because the shade cloth itself absorbs a tremendous amount of heat, it can actually melt if no air flows between it and the glazing surface. When shade cloth is used over synthetic glazing (or glass, for that matter) that is set on an angle, allow a ¾ to 1 inch (1.9 to 2.5 cm) air space between the glazing and the shade cloth. This can be accomplished if the external glazing stops or trim that secure the glazing are thick enough to hold the shade cloth off the glazing when the shade cloth is tightly stretched and held in place with a weighted rod, a clamping system, or shock cords.

Shade cloth is also very useful as a combined sunshade/insect screen when installed on the outside of a conventional window, on the inside of a casement or awning window, or under an operable skylight or roof window. To use shade cloth for this purpose, simply remove the conventional screen from its frame and install shade cloth in its place. For a skylight or roof window that does not already have a screen accessory, the screen can be made by cutting a piece of shade cloth to the appropriate size and attaching it with magnetic tape, Velcro strips, or Poly Fastener (see Appendix).

Another possibility for external shading is a screening material that has small (about 1 inch [2.5 cm]) louvers built into it. Louvered screens are manufactured by Kaiser Aluminum and KoolShade Corporation (see Appendix). Since these screens are usually mounted in frames, the cost of frames and installation increases the total cost of this option for shading your solarspace. Louvered screens permit a relatively unobstructed view and are quite durable. Depending on the specific product you select, the louvers are set at an angle of 25 to 40 degrees from horizontal and are designed to block all direct sunlight when the sun is higher in the sky than the predetermined angle of the louvers. Louvered screens are designed to be used vertically, so if you consider using them over sloped glazing you must add the number of degrees the glazing is set from vertical to determine the critical angle of the louvers. For example, glazing sloped at 60 degrees from horizontal is 30 degrees less than vertical (90 degrees). Therefore, screens with louvers set at 40 degrees, when used on glazing sloped at 60 degrees, would admit direct sunlight until the sun reached 70 degrees above the horizon (30 + 40 = 70)—which at some latitudes is the noon altitude of the sun throughout the cooling season. If you use louvered screens or sloped glazing, be sure you know at what angle the screen will block the sun. All of this addition, sun angle calculation, and sloped glazing speculation aside, be assured that louvered screens are very effective for shading vertical glazing in summer, and that these screens must be removed during the heating season.

As previously stated, shading devices placed on the outside of the glazing are more effective than internal shading devices for reducing heat gain to the interior of a solarspace during the cooling season. Although an insulating or radiant curtain or shade on the inside of the glazing will block the sun from striking interior surfaces, it will allow heat to build up between the glazing and the curtain, thereby increasing the temperature in the solarspace. This is especially true of a curtain that is tightly sealed around the edges since hot air cannot escape from behind it. Does this mean that an insulating or radiant curtain or shade that is designed to reduce heat *loss* in winter cannot also be effectively used to reduce heat *gain* in summer? Yes and no. It can, *if* the top and bottom tracks and edge seals can be loosened to allow airflow between the curtain and the glazing, and also *if*—and only if—roof vents or skylights in the insulated portion of the solarspace roof allow hot air to escape. It cannot if the tracks and seals cannot be loosened to allow airflow, and if there are no roof vents or other openings in the insulated portion of the solarspace roof. As an example of this principle, consider the PSDS radiant curtain illustrated in figure 6–7. During the heating season the upper and lower tracks of this curtain are fastened tightly to the glazing framework with screws to *prevent* updrafts behind the curtain. The screws holding the track in place can, however, be "backed out" during the cooling season, and the magnetic edge seals left unattached, to *encourage* an updraft that would ultimately carry the heated air through openings in the insulated portion of the solarspace roof. Wonderful! . . . but this scheme has some drawbacks. Alumi-

nized films and some types of aluminized fabrics are opaque, so if you have plants in your solarspace, they will need to rely on incoming light from east and west windows and doors, or shaded skylights, or both. Another possible drawback is that some types of curtain fabrics and backing materials will fade and otherwise deteriorate in direct sunlight over extended periods of time.

## Venting Strategies for Summer Cooling

If your solarspace is well shaded and has adequately sized high and low vents, you should have no difficulty keeping the interior comfortably cool in summer—except perhaps during record-breaking heat waves. Basically, all you need to do is keep the vents open between the solarspace and the outdoors. Because the solarspace and the house enjoy an intimate relationship, however, the interaction between the two must also be considered.

Undoubtedly you will have to experiment a bit to find out what will work best given the ventilation pattern of your house and solarspace in relation to prevailing summer breezes, temperature, and humidity. For example, you might try leaving the vents, windows, and doors between the solarspace and house open during the night and early morning, then close them during the hotter parts of the day. This strategy will work best in dry climates or where there is a marked difference between daytime and nighttime temperatures. Where this is the case, the mass inside the house and solarspace will cool down at night and give off the excess heat and moisture it absorbed during the day.

In more humid climates, where the temperature differs only slightly during the day and night, or where no nighttime breezes blow, the strategy of opening the house to the solarspace at night and closing it during the day will probably not work well in the hottest parts of the summer, although it may work in late spring and early fall when nighttime temperatures are lower. Again, the best thing to do is try several strategies and see what works best in relation to the vents of your solarspace and the prevailing weather conditions. Isolating the house from the solarspace during the day—while leaving the vents between the solarspace and the outdoors open—is a good starting point. You may find it necessary to modify this strategy somewhat if your solarspace covers the entire south wall of your house and the prevailing breezes are from a southerly direction. For example, you may want to have the vents to the house open to allow air entering the solarspace to flow through the house.

If you have followed the advice offered in chapter 3 and left a space in one of the end walls or in the roof of your solarspace to add an electric exhaust fan or power roof vents, try out your natural ventilation plan for at least one cooling season before you add the electric fans. If you want to increase the effectiveness of the wind turbines in the roof of your solarspace, you could either raise the height of the stacks on which the turbines are installed, install a thermostatically controlled circular duct fan inside the turbine stack, or do both. Either approach will increase airflow, but try increasing the stack height first.

Whenever you have the option of improving airflow by natural means, you will avoid expenses for the purchase and operation of fans—as well as avoid mechanical noise pollution—if you opt for a natural approach. By the same token, more external shading of the solarspace glazing areas may be preferable to adding electric fans since the increased shade could potentially reduce the temperature inside the solarspace and thereby eliminate the need for fans.

The level of comfort for people (and plants) who inhabit your solarspace in the summer will be affected by the rate of airflow and the level of humidity in the space. Generally, the faster the air moves, the more comfortable you will feel, even when the temperature is relatively high. Although air movement assists your body's evaporative cooling system, if air moving across you is high in humidity, you are not likely to feel too cool.

If the summers are hot and humid where you live, you may find it necessary to develop a strategy to reduce the humidity of the air entering your solarspace and house in summer. As suggested above, interior masonry surfaces naturally absorb moisture. With adequate airflow you should have no problems with condensation on these surfaces, and the masonry will therefore serve as a sort of dehumidifier/evaporative cooler. The addition of an air-to-air heat exchanger to the solarspace will also help reduce humidity year-round and will provide people and plants with fresh air.

The humidity of incoming air can be reduced by placing trays filled with a dehumidfying substance such as silica gel where air entering the solarspace will pass over or through it. Silica gel or other desiccant material can sometimes be purchased in large quantities from glass manufacturers who use it inside the aluminum frames of double-insulated glass units to reduce fogging between the two panes. It is also available from chemical suppliers.

Desiccant material should be placed in trays inside the solarspace vents and must be covered with a fine mesh like mosquito netting to keep it from blowing away. Some desiccants are made from calcium oxide or sulfuric acid—neither of which you would want in your house—so be sure you know what you are buying. Silica gel is probably safe to use, but if you want to be absolutely safe, use white rice. When the desiccant material becomes saturated with moisture, it can be placed outdoors in the sunlight to dry out again. Silica gel or rice can be dried in an oven set at the lowest temperature for thirty minutes to an hour.

## MAINTENANCE

After you finish building your solarspace, very little work is required to keep it operating efficiently for heat production. The glazing surfaces must be kept clean inside and out, especially during the heating season. Window cleaning and a yearly once-over to replace caulking is really all that is absolutely necessary. As is the case with the rest of your house, if your need for spotlessness and order is high, you may clean, dust, scrub, polish, and rearrange the solarspace to your heart's content; if it is low, you will not have to do very much housekeeping to keep it in satisfactory order.

You will undoubtedly find that cleaning your solarspace glazing is much easier if you invest in some professional window-washing equipment. A brush and a squeegee with a long, interchangeable handle and some solvent purchased from a janitorial supply firm will come in very handy for cleaning large expanses of glass. Dirt and dust on synthetic glazing can simply be hosed off. If you prefer not to use chemical window-washing solvents and other paraphernalia to clean the glass of your solarspace, wash the glass with white vinegar or ammonia and water and dry it with newspaper.

Repainting or staining the exposed framework of your solarspace may be necessary from time to time. And, although it will probably not affect the performance of your solarspace, it may help prevent wood from rotting because of high moisture conditions. Some types of synthetic glazing may need replacing every few years or recoating to retain light transmittance at high levels. Check with the glazing manufacturer.

Other maintenance tasks that may apply to your solarspace include draining water preheat systems yearly to check for rust or corrosion, and checking the alignment of fan blades. It may also be necessary to lubricate fan motors and bearings according to manufacturers' recommendations.

If you are going to use your solarspace for growing houseplants or food, other important tasks, such as changing the soil in growing beds, should be performed. These are described in the next section.

# GROWING PLANTS AND FOOD IN YOUR SOLARSPACE: A STARTER COURSE

If you are the kind of gardener whose only past success is with growing something green and fuzzy in your refrigerator, take heart. A solarspace can be an excellent environment for growing houseplants, vegetables, fruits, herbs, maybe even something as rare and exotic as an orchid or anthurium. No matter how you have decided to use your solarspace, grow something in the sunshine. You may be surprised to discover that growing plants can add a new dimension to your home and your life. Starting with just a few plants can show how they (and you) respond to the new conditions. If you think of yourself as a "black-thumbs-gardener," start with a philodendron, snake plant, begonia, patio tomato, or geranium; any of these can do well with sunshine and a little water, even if neglected and abused. When you see how well

these plants are doing, you will no doubt be eager to try a wider selection. Do not be discouraged from growing because of a lack of experience or bad luck in previous gardening attempts. The unique environment inside a solarspace may enable the inexperienced gardener to grow and propagate plants easily, whereas the experienced gardener may find a brand-new challenge.

Remember that every solarspace is different, as it should be designed to work for a particular home. Conditions inside a solarspace are unique in design, orientation, and climate. By keeping this in mind, you can easily understand why the suggestions and recommendations presented below are intended only as a place from which to start. Only you can determine what will and will not thrive in your own solarspace.

It is easy to get carried away by the possibilities solarspace gardening offers. So, in addition to growing plants that will thrive, remember to choose plants that you like. What advantage is it to have tomatoes, zinnias, and strawberries growing like crazy in your solarspace if they give you hives or bring on an asthma attack? Start with something you like; then gradually add as many plants as you want, but be careful not to shade too much mass during the heating season if you want your solarspace to remain comfortable.

## Keeping Records

Get started right in this new adventure of solarspace gardening by keeping careful records. These records should include details of soil mixtures; watering schedules; new plants introduced; germination of seeds—giving names, sources, dates planted; propagation methods and dates; inside and outside temperatures; weather conditions; invasion of pests or diseases, and steps to control them. Any information that could be of benefit in drawing conclusions will be very valuable in planning future gardening in the solarspace.

## Planning Your Solarspace Garden

Before the interior of the solarspace is completed, decide how you choose to garden: in ground beds, raised beds (see fig. 6–12), straw

*6-12. Container gardening on raised tables in Seawell solarspace (Photo: Darryl J. Strickler)*

bales, soil bags, or containers. If you choose to use ground beds, plan them as an integral part of the solarspace so they work with the mass and catch and use runoff from watering the plants or humidifying the space. If you are undecided at this point, you may want to plan beds that could be filled with earth, stones, brick, or wooden slats on which containers, raised beds, straw bales, or soil bags can be arranged now, allowing space for ground beds at a later time.

A number of factors must be considered when choosing your gardening method.

■ The soil in which the plants are to grow should be changed about once a year. Concentrated growing quickly depletes the soil of essential nutrients that should be replaced after each harvesting. In an outdoor garden, nutrients may be added year after year, but within a finite growing area like your solarspace, some or all soil must be removed to be able to change the soil or add additional nutrients. If you are gardening in ground beds or raised beds, changing the soil will involve digging out and filling in; straw bales and soil bags can be replaced; containers can be taken out individually, dumped, and repotted as needed.

■ Plant roots are less restricted in beds and

will not become overcrowded as quickly as they would in containers.

■ Staking or supporting climbing or trailing plants is easier in beds than in containers.

■ Gardening in a bed—whether on the ground or on raised tables—may be more enjoyable to some, as it is more like working in an outdoor garden.

■ Plants in containers can be brought into the house for color and atmosphere or moved outdoors when weather permits.

■ Isolation for disease or pests is easier when plants are in containers, straw bales, or soil bags, allowing the immediate removal of unhealthy plants *before* the entire environment is affected.

## Soil Mixes

Two basic types of soil are used in solarspace gardening: a medium to germinate seeds and a general potting soil for transplanting and repotting. The germinating medium should be light, loose, and screened, because the delicate new roots need to grow with the least resistance from the soil. A mixture of one part sphagnum moss or peat moss to one part sharp sand makes a good, lightweight medium for germination, as well as for propagation from stem and leaf cuttings. (Sharp sand is used in making masonry mortar; rounded sand is the common sand rolled by water action, from rivers, lakes and oceans.) The nutrient content of germinating soil is usually not too critical because each seed contains the correct combinations and ratios of nutrients sufficient to sprout and begin growth. Sterilization of the germinating medium is important to kill most active fungi or diseases, to keep it from sprouting weeds, and to prevent damping-off, a fungus disease to which seedlings are especially susceptible. Damping-off causes otherwise healthy seedlings to shrivel and fall.

The composition of potting soil should vary with the types of plants to be grown. Here are some simple all-purpose mixtures.

*All-purpose mix 1*
1 part compost
1 part peat moss or vermiculite
2 parts sharp sand or 1 part sand and 1 part perlite
2 parts garden loam or topsoil

*All-purpose mix 2*
1 part peat moss or vermiculite
½ part compost
1 part sharp sand or perlite
1 part garden loam or topsoil

For potting cacti or most other succulents, prepare all-purpose mix 2 with an additional half part perlite or crushed potsherds (broken clay pot pieces). To this mixture add powdered (sterile or steamed) bone meal at a rate of one part bone meal to sixty-four parts mixture. Add lime in the same one-to-sixty-four ratio. When mixing soil for moisture-loving plants like fern, one part additional peat moss or compost may be added to all-purpose mix 1. This soil mixture may be enriched with powdered (sterile or steamed) bone meal at a rate of one part bone meal to sixty-four parts soil mixture.

Large plastic containers with tight-fitting lids are great for storing soil mixes and ingredients. (Many fast-food stores get pickles or mustard shipped in them.) If your growing space is limited, large coffee cans with plastic snap-on lids work very well.

Soil acidity or alkalinity is measured by pH: pH 7 is neutral, less than pH 7 is acidic, more than pH 7 is alkaline. Most plants prefer the soil slightly acidic—about pH 6.0 to pH 6.9 being neither very acid nor alkali. The pH balance is important for good plant growth and production but is usually not a problem if you are using all or mostly organic ingredients in the soil mix. A great deal of information is available about soil pH balance, checking pH factors with litmus paper or soil test kits, and correcting soil imbalances (See Bibliography). The simplest indicator may be the plants themselves.

If soil mixtures contain plenty of organic matter in the form of rotted manure, leaf mold, organic peat moss, or especially, compost, the pH is likely to be near perfect. If plant leaves are browning or look burned or if the plant is showing signs of very slow, weak growth, have the soil tested. Call your local state agricultural extension service or get a test kit and test it yourself. One of the easiest remedies for pH problems in the soil, whether to correct for too much acid or too much alkaline, is compost.

### Soil Enhancers
Numerous soil additives can be used to change: pH levels; nitrogen, potassium, and phosophorus levels; trace mineral elements; texture; and moisture retention; to name a few. Basically, the main soil enhancers are either nutrient builders (fertilizers), or soil texturizers. Many fertilizers are described in number ratios, such as 10–10–10, that refer to the content of nitrogen (N), phosphorus (P), and potassium (K), although a good, balanced fertilizer contains many trace elements as well. Manufactured fertilizers have some advantages in that they are neat, easy to use, and require little or no mixing. They are readily available in most places where plants are sold—even in grocery stores. A disadvantage of synthetic chemical fertilizers is their composition: manufactured nutrients are usually composed of the NPK elements only; trace elements must be added and balanced separately. Another disadvantage is the high concentration of the manufactured nutrients: inadequate mixing into the soil or addition of too much fertilizer can cause burned leaves or plants. Finally, even though the manufactured nutrients are strong enough to burn plants, they are quickly leached out of the soil and washed away with waterings.

Fish emulsion is an effective organic fertilizer that is diluted with water and used at regular intervals instead of plain water. Fish emulsion is a very good source of nutrients for houseplants and vegetables, and it is easily applied. It does, however, have one disadvantage: your solarspace (or your whole house) may smell like a fish market on the day you fertilize the plants. Some garden supply centers sell deodorized fish emulsion capsules that can be mixed into the soil.

Well-rotted manure is rich in nutrients and may be worked into the soil or sprinkled over the top of the soil to make a "manure tea" every time the plants are watered. Most animal manure will work; horse, cattle, and sheep manure is usually the easiest to obtain. The important thing to remember is to use only well-rotted manure; otherwise, the plants may burn and your solarspace may smell like a barnyard.

Soil texturizers make the soil light, crumbly, and loose. They allow the soil to drain better while retaining moisture. Some additives to improve lightness and drainage are sharp sand, sandy loam, and perlite. Peat moss helps retain moisture while making the soil more crumbly.

Sphagnum moss and vermiculite also help retain moisture.

### Compost

Probably the best all-around fertilizer and soil conditioner for the solarspace gardener (or outdoor gardener) is compost. Those gardeners who use it swear by it. Perhaps one reason they do is because compost is effective for so many purposes. A good compost mixture can balance soil pH by neutralizing an overabundance of acid or alkaline; it fertilizes by containing not only the major nutrients (NPK), but the trace elements as well; nutrients are released gradually—just as the plant needs them; it provides texture to the soil by making it loose and crumbly, allowing aeration to the soil; it retains moisture; and it gives off warmth to the soil and carbon dioxide ($CO_2$) to the plants.

The only real disadvantage of compost is its availability. With the exception of a very few gardening shops, compost is not sold commercially. Compost can, however, be made at home. All it takes is a little space, soil, and organic material. (If you have access to animal manure, you are even further ahead.) The location of the compost heap can be inside the solarspace or outdoors. Sacrificing a little growing space for composting will certainly pay off.

Small-scale solarspace composting has several advantages. Activity of the microorganisms that feed on the organic matter produces $CO_2$ and heat (to 150°F [65.6°C] in a well-stocked compost pile). Both are important to plants and the solarspace.

As in gardening, your compost pile will work efficiently if you remember to keep things simple. A garbage can or barrel will do nicely for composting in a limited space. If your solarspace has enough room, you could build a larger bin. The container needs ample holes; in the side for aeration and in the bottom for the drainage of excess liquids. Animal manure (if you do not raise animals, you can obtain it from a gardening store; if you are ambitious, check your local farm or zoo) and plant matter make a good compost. Coffee grounds, kitchen scraps, and wood ashes will make compost too, but if that is all you have available, additional phosphorus is necessary to balance the "stew." A well-stocked garden center or nursery carries ingredients to balance out such a compost.

Composting is simple. Start with a thin layer of soil then add any vegetable scraps, undiseased plant wastes, leaves, grass clippings, coffee grounds, anything organic you may have. Over the organic material, throw a scoop or shoveful of rotted manure if you have it, then add a thin layer of soil. Repeat the layering as you acquire more organic wastes, always finishing with a layer of soil on top to check odor. Check the pile for moisture; sprinkle it when it begins to dry out. The idea is to keep it moist but not wet; too much water will cool down the pile and slow down the composting process. Every two or three days the pile should be poked and stirred with a pitchfork or hand cultivator; frequent mixing supplies oxygen to the pile and helps release $CO_2$ for the benefit of the plants. Depending on the composition and care of the compost, a bin could be ready in as early as four weeks. If materials that are slow to decompose were used, it could take as long as eight weeks before the product is ready.

Getting the rich store of nutrients in the compost to the plants is also very simple. Make it an integral part of the soil mixture when potting or repotting plants (see soil mixes). Existing plants can be fortified by adding a topdressing of screened compost, about ¾ inch (1.9 cm), on top of the existing soil. If there is no room in the pot, make room by gently scooping away soil around the edges and top of the pot with a spoon, being careful not to disturb the roots. By topdressing, the nutrients in the compost will be released with each watering, and the compost will help mulch the plant and keep it moist.

## Beginning to Grow

With care in planting—using adequate soil mixes and sufficient fertilization—you may be able to grow not only vegetables and herbs with ease, but have continuously blooming flowers for your own home and plenty to share with friends. As you learn how easy it is to propagate and germinate plants in your solarspace, you may even think of selling plant starts to a nursery or plant shop. It is really exciting to experiment and discover what new plants can be grown and how beautifully they can be grown. Use your imagination—be creative.

The list of plants that could thrive in solarspace conditions would take a book in itself (in fact, many such books are listed in the Bibliography). Some varieties of a specific plant may do better than others. Plants and seeds of flowers and especially vegetables that are acclimated to more southern latitudes will often produce better in solarspace conditions than will northern-adapted strains. A lot depends on the climate you will be maintaining inside the solarspace, as well as to what degree your solarspace is integrated with your house.

### Houseplants

Almost any solarspace can be dressed up with greenery. Some simple-to-grow, inexpensive plants follow.

- snake plant or mother-in-law's tongue (*Sansevieria trifasciata*): use all-purpose soil mix 1 or 2; water occasionally.
- spider or airplane plant (*Chlorophytum comosum*): use all-purpose soil mix 1 or 2; water occasionally. This plant is beautiful in a pot or hanging basket. It will not send out runners until it is pot bound, so do not repot until absolutely necessary, and even then, do not use an overly large pot.
- wandering Jew (*Tradescantia albiflora* or *Zebrina pendula*): use all-purpose soil mix 1 or 2; water occasionally or when it wilts. This plant is beautiful in hanging baskets. Pinch the tips weekly to grow a compact, lush plant.
- philodendron (*Philodendron cordatum* or *P. scandens oxycardium*): use all-purpose soil mix 1 or 2; water occasionally. This plant will trail and vine if you allow it. Pinch branch ends frequently for a full, compact plant.

The above four plants are easy to care for, requiring only occasional watering. They are all easily obtained from plant shop, supermarket, or variety store. You could also get "starts" from a neighbor, friend, or your granny. All except the snake plant root easily in water.

### Colorful Plants

Now, how about a little color? A begonia (*Begonia* species) or geranium (*Pelargonium* species) is easy to grow and will provide clouds of beautiful color. Be sure you choose a bloom-ing variety. *Begonia semperflorens* and *Pelargonium hortorum* are two common varieties that are prolific bloomers. Use all-purpose soil mix 1 and give the plant extra compost, manure tea, or liquid fertilizer every few weeks. Make sure the soil is well drained (potsherds in the bottom of the pot will help), and allow the soil to dry before watering again. Either plant may be taken outdoors in spring and returned to the solarspace in the fall. If the plant has become weather battered, sunburned, or has outgrown its space indoors, simply cut off some healthy-looking stems (be sure to have four or five stem joints), cut off any leaves on the bottom two joints, and root in water. If there are any blossoms on the stem, remove them, to focus the plant's energy on developing roots. When the root system is thick and strong, the new plant may be planted in soil. By starting new plants and providing them with plenty of good soil, sunshine, and water, it is possible to keep begonias or geraniums blooming continually.

Another hardy plant profuse in color is coleus (*Coleus blumei*). The foliage of the coleus, not the flowers, gives the plant its beauty. It requires the same basic care as geranium or begonia and can be taken outdoors and repotted from cuttings.

Petunias (*Petunia hybrida*) and marigolds (*Tagetes* species) do very well planted from seed or seedlings in all-purpose soil mix 1 or 2. Each likes well-drained soil that is dry between waterings (to keep lots of blooms, do not allow it to be dry too long). For winter color sow from early fall to midfall. Petunias provide abundant color and a delightful fragrance. Marigolds add rich, sunny color and are effective in repelling insects, making them useful for interplanting with vegetables or other plants.

Nasturtiums (*Tropaeolum* species) will bloom profusely in sandy, dry soil. Plant in all-purpose mix 2 and allow the soil a chance to dry between waterings. Different varieties of this plant are very versatile in form: some become bushes in pots or beds; others vine up walls or trellises or trail over hanging pots. Not only does this plant produce beautiful yellow and red flowers, but the tender leaves, flowers, and buds are edible. The leaves and flowers add a delicate peppery flavor to salads and can be used as garnishes; the flower buds can be pickled for use as capers. Nasturtiums will grow and bloom year-round with continuous plantings. (Starting seeds every three or four months should provide constant growth and blooming, but the frequency will have to be adjusted to your own solarspace.) Besides growing inside, the vining variety makes a good outdoor sunscreen for the cooling season when trained over the south glazing. When growing nasturtiums for eating, provide them with a richer soil mixture (try adding extra compost or use all-purpose soil mix 1), and water more frequently. If you think these virtues are enough for the humble nasturtium, wait, there is one more. It is also delightfully fragrant.

Other hardy bloomers for the solarspace include zinnias (*Zinnia elegans*), which do well in hot, dry conditions with regular waterings; pot marigolds (*Calendula officinalis*) and mums (*Chrysanthemum morifolium*), both of which prefer cooler areas near the glazing or floor for best flowering.

## Fruits and Vegetables

By now, the solarspace is alive with greenery and color. The neighbors are amazed at your talents, and your Aunt Harriet thinks you are paying a plant rental agency to care for the lush plants in your solarspace. If you were doubtful before, you should now have the confidence to grow some food. Maybe you would like to start with tomatoes (*Lycopersicon esculentum*). (If you do not like tomatoes, try some leaf lettuce or spinach.) Patio-variety tomatoes are good to start with because they grow compact and sturdy and they usually self-pollinate. They are good producers and can be planted in beds or containers. Plant them in rich, well-drained soil, adding an extra handful of compost or peat moss. Let the soil dry briefly, between thorough waterings; give the plant plenty of sun and watch it grow. Very soon, yellow star-shaped flowers will start blooming; each blossom is a potential fruit. When the little green marble-sized tomatoes form, pinch the smaller ones off to allow the plant energy to produce a few larger, better-tasting fruits. The plant will continue flowering and producing for a long time. For best results feed it frequently, mulch with compost or leaves, water daily if needed, and keep it pruned to a manageable size. After tasting a vine-ripe, juicy, red tomato in mid-

January, you will likely become a real solarspace gardening enthusiast.

Other edible crops for beginners are leaf lettuce (*Lactuca sativa*); spinach (*Spinacia oleracea*) (New Zealand variety is a winner in most solarspace gardens); cucumbers (*Cucumis sativus*); squash (*Cucurbita* species) (zucchini and yellow squash do well; fruits are easier to support if vines are trellised). Radishes (*Raphanus sativus*) and peas with edible pods (*Pisum sativum*) are not only popular solarspace crops but are very easily grown. Strawberries (*Fragaria* species) also do very well and can make attractive hanging baskets or strawberry jars.

If your solarspace tends to have temperatures in the lower end of the ideal range of 50° to 85°F (10° to 29°C), you may want to grow oriental vegetables. One easily grown oriental is Chinese cabbage (*Brassica pekinensis*). Other cool-weather crops include broccoli (*Brassica oleracea italica*), brussels sprouts (*Brassica oleracea gemmifera*) and cabbage (*Brassica oleracea capitata*). Cultivation on or near the floor and under south windows will provide the cooler temperatures these plants need.

Herbs make a rich, sometimes fragrant addition to the solarspace. Almost any herb will do well in the solarspace with regular watering and occasional feeding. The central area of solarspace, where there is moderate temperatures and light, is a good area for cultivation. Many herbs are good at repelling insects; some enhance the flavor and health of other plants. Interplanting herbs—or intercropping—with other plants may help promote growth (see *Pests and Plant Diseases* later in this chapter). You can also sell herbs to, for example, plant shops, fine restaurants, and specialty grocers.

For those with space and ambition, dwarf and extradwarf fruit trees can be grown in containers. Even grapes and bush berries can be grown in a solarspace under the right conditions.

Dwarf and extradwarf varieties of trees should be kept in containers to promote dwarfing. Prune thoroughly and regularly to produce a sturdy, compact, productive tree. These trees may also require unpotting and trimming of the roots every few years. Full or nearly full-sized fruits can be produced by regularly thinning smaller growth.

It is important to check pollination requirements when buying dwarf and extradwarf trees. Some trees will cross-pollinate well with each other if two or more trees are planted; others require trees of another variety to do well. Unless your solarspace is large enough for a small orchard, select varieties of trees that are self-pollinating. Even a self-pollinator may need assistance. (Tips on pollination are included below.)

Trees that are good producers and also make good ornamentals, include dwarf or extradwarf apple, peach, nectarine, apricot, plum, fig, orange, lemon, lime, and mandarin orange.

Some gardeners feel nursery stock that is grown in southern climates adapts somewhat easier to solarspace conditions than the stock grown in northern areas. (See Bibliography for more information on growing small-sized fruit trees, grapes, and berries.)

## Starting New Plants

Once you have wet your feet (or gotten fertile soil under your fingernails), you will probably look forward to trying many new plants in your solarspace. The conditions in your solarspace may not be correct for growing absolutely everything you can think of, but then, that is part of the challenge. Aside from the limitation of your indoor climate, you are limited in what you can grow only by your imagination. So go ahead, order all the plant and seed catalogs you like, and turn those daydreams of blossoms and ferns and vegetables into the real thing.

Opening the first page of a new seed catalog is opening a door to another dimension of your solarspace gardening. With the acquisition of a few seed packets, it is possible to turn that solarspace into an absolute jungle. (It is a virtue to know when enough is *enough*.) Learning to germinate seeds and propagate plants is like taking off the training wheels—it may take a few mistakes to get the hang of it. This is where the record keeping of all work in the growing space really pays off. Germinating and propagating will allow you to share or trade plants with friends or start plants you have wanted but could not buy.

The soil for germination, as described in the earlier section on soil mixes, should be light and free of stones, twigs, and other debris. Of course, you may purchase peat pots, cubes, or pellets for use in germination or propagation, but plenty of throw-away containers are more than adequate. Cottage cheese containers, margarine tubs, fiber milk cartons (cut off the top or one side), plastic foam coffee cups or fast-food sandwich boxes, and aluminum frozen-food loaf pans are a few such containers. Be sure to make good drainage holes in the bottom, fill three-fourths full of germination medium, and you will be ready for planting.

### Stem and Leaf Cuttings

Propagation of stem or leaf cuttings is easily done by burying about ½ inch (1.3 cm) of the cutting into the medium and watering. The spider plant (runners), wandering Jew (stem cuttings of tips with four or five joints; the two leaves closest to the cut removed), and philodendron (stem cutting; same as wandering Jew) can all be rooted in germination medium that is kept moist but not wet. In fact, these can also be rooted in a glass of tap water.

African violets (*Saintpaulia ionanthas*) can also be easily propagated from a leaf cutting. Snip off a strong, healthy outer leaf, being sure to cut near the plant. Plunge the leaf into the germination medium (dip leaf in a rooting powder first if you like) and water it. Keep the medium moist, and soon tiny round leaves will appear around the stem. When leaves are the size of your little fingernail, transplant into an African violet mixture (one part sphagnum moss, one part vermiculite or perlite, two parts compost) and keep it moist; feed with compost or African violet food.

### Starting Seeds

When planting seeds, prepare containers and fill them approximately three-fourths full with germinating medium. Water gently so the surface is not splattered or pitted, but the medium is soaked. Sprinkle seeds lightly over the surface and gently touch them with your fingertips to firm seeds into the medium. A very light dusting of screened soil will help hold the seeds in place. Larger seeds may be lightly covered and gently firmed with fingertips. The germinating medium provides texture for the developing root system to grow into.

Very tiny seeds, such as those for lettuce, leeks, carrots, mustard, and petunias, may produce seedlings so small and delicate they are difficult to transplant. Moreover, plants in which the root forms the edible portion, such as carrots and radishes, should not be transplanted. Sowing seeds directly into beds, containers, or peat pots will eliminate the need for transplanting. Scoop out approximately ½ inch (1.3 cm) of potting soil in the final bed or container and replace with screened germinating medium. Then plant the seeds as described above.

Moisture and warmth are very important to successful germination. To keep seeds and seedlings free of drafts and in controlled humidity, enclose the germination container in a plastic bag and seal it. Open the bag once a day for a few minutes to check the soil moisture and exchange some fresh air. (You might even blow into the bag to be sure there is enough $CO_2$ for growth.)

Since many seeds require higher temperatures for germination than for growth, soil temperatures are important. When trying to germinate during cold weather, it may be necessary to provide additional warmth to the seeds—especially if your solarspace is uninsulated from night sky radiation. A very inexpensive way to heat seeds is to run thermostatic heat tape under the containers. (Heat tape is available in some gardening or seed catalogs. Heat tape is also used on trailer and camper water supply lines and garage or outdoor water supply lines.) To distribute heat evenly, run tape back and forth across the bottom of a growing flat and cover with a little sand; set germinating containers on the sand. *Be sure* to choose a heat tape designed to be used outdoors. It should be labeled as waterproof—otherwise, when watered, your plants may get a real charge from the experience and so could you.

If your solarspace sports water drums or other added mass, you could make a shelf on the top for the germination nursery. The slow release of warmth from the mass will provide sufficient heat to encourage sprouting. High shelves on the mass wall are another good place for sprouting. (If you make your own bean sprouts for cooking, a shelf like this is perfect for your sprouting container. Rinse the sprouts often, because they can spoil quickly.)

As the new seeds begin sprouting, you will notice one or two leaves that are first to break the soil. These first leaves are called seedling leaves. The next set of leaves are called true leaves and are usually similar in appearance to those of the adult plant. When the seedling has produced at least one or two sets of true leaves, it can be fed a fertilizer tea of either a diluted fish emulsion preparation or a composted manure or organic fertilizer at a rate of one part fertilizer to sixty-four parts water. As the second or third set of true leaves appear, the seedling may be transplanted to an intermediate container. The intermediate planting is not essential but allows continuing growth and nurturing without depleting the soil in the proposed planting area.

Any plants being started in the solarspace for use outdoors should be hardened off before the actual transplant. All this amounts to is taking the seedlings outdoors at least one week before transplanting. Give the seedlings shelter on a covered porch or under a tree to protect them from the wind or sudden chills. Check seedlings daily for drying out and sprinkle with a hose or watering can if necessary. If frost is forecast, take them indoors.

## Placement of Plants

As seedlings become ready to transplant into a growing space, it is time to give some consideration to where the plants should grow. The solarspace designed primarily for heating and cooling has two very distinctive seasonal climates, by the nature of its function and design. The winter climate is intended for heat production and the solarspace will be very warm, with sunlight on the floor, maybe even on the back wall; short hours of sunshine; and cooler temperatures at night (unless the windows are protected from night sky radiation). The summer climate, however, will be somewhat cooler and have a little lower light (because of venting and shading) even though the hours of sunlight increase.

Certain areas inside the solarspace will provide optimum light and temperature conditions for certain kinds of plants. Being aware of these subclimates will help you place each plant in the location best suited for it.

Air temperature is coolest on the floor and warmest near the ceiling. Temperature is also cool next to the outside glazing and warm next to the inside wall. Thus, the coolest space is near the floor, under the southside glazing; the warmest space is against the back wall, near the ceiling.

Light is brighter next to the south glazing and dimmer next to the back wall. It is also bright near the floor and dim near the ceiling. The brightest light is therefore near the floor under the south glazing; the lowest light is near the ceiling, against the back wall.

It is possible to locate plants in the subclimate in which they will de best. For example, beets, broccoli, radishes, spinach, lettuce, coleus, pansies, cyclamen, and bromeliads grow best in a cool, sunny area. All of these plants can grow well in winter on the floor, in containers or beds near the south glazing.

Close to the back wall, where the temperatures are moderately warm and there are fewer hours of direct sunlight, impatiens, petunias, mountain or coastal fern, rubber trees, leafy green vegetables, strawberries and other berries, peas and beans, and oriental vegetables grow very well.

Tomatoes, melons, cucumbers, hibiscus, gardenia, zinnias, and some tropical and flowering plants enjoy the warm, sunny space above the floor in front of the south windows. The climbing traits of some of these plants can be used by planting in beds or containers on the floor and trellising the vines in front of the windows (see fig. 6–13). Be very aware of what is being shaded by any plant, because too much mass can easily be shaded, decreasing the heat storage during a major portion of the heating season.

A perfect place for most hanging basket planters—especially bloomers—is from the ceiling in front of the windows (see fig. 6–14). Petunias and begonias like this area and make beautiful hanging baskets. Herbs also like this area. One exception to this is the fern family; tending to prefer filtered light, they may do better hanging near the back wall out of direct light. Other plants that make beautiful hanging baskets are geraniums, lantana, purple passion or velvet plant, philodendron, spider or airplane plant, Swedish ivy, grape ivy, wandering Jew, leaf lettuce, spinach, and strawberries.

Many tropical plants that like warmth and low light will do well near the ceiling and back wall. Most plants that like warm temperatures and low or indirect light also need higher humidity (think of those rich, lush jungle scenes in the movies). To maintain high humidity, keep the soil rich in compost or peat moss. Mist the plants' stems and leaves frequently. Setting

6–13. Tomatoes on trellises behind south glazing in Mitchell solarspace; note misting system (Photo: Darryl J. Strickler)

6–14. Spider plants, wandering Jew, and begonia in ideal location for hanging plants (Photo: Darryl J. Strickler)

pots on trays lined with pebbles and filled with water will help. Hanging wet laundry to dry in the solarspace will also add moisture to the air. Plants that can thrive in this solarspace location include African violets, gloxinias, orchids, and ferns. This warm, lush area of the solarspace has equally favorable conditions for breeding molds, fungi, and other diseases, so air movement and careful observation are important.

If dwarf fruit trees, grapes, or berry bushes are planned, try them against the east or west side of the solarspace. Here they should still get ample light, but their foliage will not shade too much mass. It is a good idea to keep these plants in containers, at least until you have found the location where they like to grow.

## Pests and Plant Diseases

The solarspace can grow strong, healthy plants if you keep a careful eye out for pests and disease. Quick and positive identification will aid in choosing a suitable control that will check the problem before it becomes epidemic. Disease and infestation can ruin an outdoor garden within a period of one week, but in the select environment such as a solarspace, destruction could happen within a matter of days—every hour counts. Know the enemy. Keep an inexpensive pocket field identification guidebook handy for quick reference (see Bibliography). A magnifying glass is very helpful in discovering the tiny invaders.

You can easily check plants regularly while watering. Careful inspection of leaf condition and color, any flowers, stems, and the top of the soil around the plant could reveal cutworms, slugs, fungus, root rot, mealy bugs, and other problems.

### Prevention

The best tactic against pests and disease is prevention. Keep everything in the solarspace as clean and uncluttered as possible. Remove all dead plants, pruned parts, and empty pots from the area. Do not leave tools, watering hoses, empty flats, or other odds and ends stacked around. Any of these items could harbor a mouse or a cricket, at the very least. Keep tools properly stored, walkways cleared, workspaces clean. Dispose of any infested or diseased (especially virus and fungus) plant parts by burning, unless the plant suffers from tobacco mosaic virus.

Tobacco mosaic virus is transmitted by smoke. That is why plant parts with tobacco mosaic virus should not be burned but buried deeply in a corner of the yard. In addition, smoking should be strictly prohibited in the solarspace when plants are present. Tobacco mosaic virus causes leaves to curl and become deformed and causes stems to wither. To-

matoes, melons, squash, and cucumbers are very susceptible, as are some ornamental plants.

Ventilation is a good preventive measure for keeping the growing area healthy. If the solarspace is isolated or closed off during a particular season, try using a small table fan to move air among the plants. It is difficult for problems to take hold when plants are well ventilated. Using a fan can greatly reduce or entirely prevent mildew. A fan may also be used to help in pollination when plants are in bloom. Moreover, ventilation is important in bringing more $CO_2$ to the plants from the rest of the house.

All new plants should be quarantined before being introduced into the solarspace. Keep them in the garage, kitchen or back bedroom—any place away from the solarspace—until you are sure the plant is not carrying pests or disease.

The earlier you start preventive measures, the better your chance of protecting your plants. You can start with the potting soil.

Before bringing a new batch of potting soil into the solarspace, sterilize or pasteurize it. Some commercial potting soil has already been sterilized and will be labeled as such. Pasteurizing or sterilizing kills most pests and their eggs, weed seeds, and other undesirables in the soil. To pasteurize soil at home, simply spread the soil to a depth of about 2 inches (45 cm) in a broiler tray or cake pan and "bake" at lowest oven heat for forty-five minutes to one hour. Store any unused soil in a plastic garbage can with a tight lid or in a sealed, heavy-duty trash bag.

For the gardener planning to start on a small scale and gradually add plants, home soil pasteurization is a worthwhile project, as it offers maximum protection from soil diseases. Obviously, for the gardener planning ground or raised beds or for the gardener planning a large-scale beginning, home pasteurization is not the answer. Neither is the expense of buying sterilized soil a likely answer. The large-scale solution is to use compost freely (it is pasteurized by the high temperatures it reaches in processing) and carefully observe for infestation.

### Insect and Disease Control

Almost as many insect and disease control methods and reasons for promoting them exist as do insect pests and diseases. Perhaps the most important consideration to remember in choosing a control method is this: the air in the solarspace will be circulating into your living space during the heating season and maybe the cooling season too. Sure, literally hundreds of products on the market repel or kill almost anything that could creep, fly, or float into your solarspace. But knowing you will be living with and breathing whatever is sprinkled or sprayed in your environment may persuade you to keep your controls as organic and natural as possible. If you do decide on a miracle bug blaster, plenty of instructions are available from manufacturers and the store or nursery selling it.

One exception to avoiding chemicals is the use of mothballs or moth crystals. They can be buried in the soil or sprinkled on top of the soil to repel most crawling pests or root pests. If you live with a cat or dog, this will usually keep them out of the beds as well.

A standard pressurized tank sprayer is a worthwhile addition to your gardening tools. If your budget prohibits investing in a tank sprayer, wash out a household-cleaner pump spray bottle. It has to be filled more often but will not be as heavy to use.

Many spraying solutions are effective on insects, some virus strains, and fungi. They can be applied separately or in combination.

An easy, all-purpose repellent is a soap solution. Make sure you use a soap—for example, Ivory, Fels-Naptha, or a Castile—as detergents could burn the plants. If using a liquid soap, mix one part liquid soap to sixty-four parts water. If using powdered soap, mix one part powdered soap to thirty-two parts water. Adjust concentration, if needed, and spray on plants, flowers, fruits, and vegetables. If insects persist, increase concentrations up to 1:48 with liquid soap and 1:24 with powdered soap. If you spray frequently (more than once a week), dilute the solution with more water (up to 1:96 for liquid soap and 1:48 for powdered soap).

Other general repellent solutions include:

- A strong tobacco tea—good for repelling most insects, including thrip, although not recommended for tomatoes, potatoes, melons, or squash. Soak any tobacco, even cigarette butts, in hot water. Strain the solution before spraying.
- A strong pepper tea. Brew the same as tobacco tea, only use cayenne, tabasco, or crushed red pepper pods.
- Garlic juice solution. The stronger you can make this and still stand to use it, the better it will work.
- Oils of pennyroyal, mint, or sassafras, separately or in combination. A few drops of soap will help emulsify these oils into water.
- An insect malt. Claims to the effectiveness of this method make it worthy of consideration, even though it appears a bit bizarre. The idea is to hand-pick as many insects as possible and blend them in a blender, into a malt. This produces a liquid that is said to repel the same kind of insect that was blended. Strain before spraying.
- Flour and buttermilk. Mix four parts flour to one part buttermilk and dilute with enough water to spray or sprinkle. This solution is a good general spray, especially effective against red spider mites.

Dusting powders are also easily made from common ingredients. Both cayenne pepper powder and tobacco powder can be dusted or sprinkled on plants, other than tomato, potato, melon, or squash, as a good insect and vermin repellent. These powders will sometimes repel dogs and cats too.

A pump or squeeze-type sprayer will make dusting easier; several types are available at a garden supply store. You can also use a large salt shaker or empty body powder or talcum containers. Just be sure to dust into branches and stems and the soil around the stems.

Crumbled tobacco leaves from dried chewing tobacco or snuff or crumbled cigarette butts sprinkled on the soil will repel crawling pests. To help prevent the introduction of tobacco mosaic virus into your solarspace, do not use tobacco in any form on tomato, potato, melon, or squash plants.

Wood ashes *lightly* sprinkled in a ring around the plant will repel cutworms and slugs.

Diatomaceous earth sprinkled or powdered around the plants is effective against worms and most soft-bodied pests.

Rubbing alcohol (isopropyl alcohol) on a cotton swab can be used to clean mealy bugs, aphids, and sometimes scales from plant stems and leaves. Scales can also be scraped off with a toothpick.

Heavily infested potted plants can be

dunked upside down in a soapy solution (the same solution described above for spraying). Use a cloth or plastic bag to cover the soil and hold it in place as you dunk the plant.

Tender plants can be protected from cutworms and other crawlers with a protective cylinder. Use such everyday items as yogurt containers (especially the waxed kind), coffee cans, milk cartons, or large frozen juice cans. Simply cut out the bottom (and top, if needed) and press the cylinder about one-fourth to one-third of its length into the soil around the seedling, being careful not to push too deep and injure shallow roots. The cylinder needs to be only large enough to encircle the stem amply; when pressed in place, it should be 2 or 3 inches tall. A disk cut from black roofing felt, plastic, or aluminum foil will also protect seedlings—especially tomatoes and peppers—from cutworms and some crawling pests. Make the diameter of the disk about the same diameter as the seedling's foliage; cut into the disk to the center, then cut a small hole that will easily fit around the plant stem. Aluminum foil is thought to repel aphids but may attract a magpie or crow to your window.

Some plants enhance or inhibit growth and production of other plants. Among the qualities of a compatible plant are characteristics of aroma, root formation, soil demands, and physical growth. Cultivating two such compatible plants together is called companion planting, or intercropping. An example of companion planting is cultivating basil next to tomatoes. The fragrance of the basil deters some insect pests of the tomato plus the basil will promote a better flavor in the tomato crop. The tomato, on the other hand, encourages a crop of flavorful basil.

All the elements that make a plant compatible can also make it incompatible. For instance, the same tomato that was compatible with basil is not compatible with kohlrabi; production and quality will be inhibited in both plants. Other incompatible qualities include both plants requiring large amounts of the same nutrient, both plants attracting the same insect pests or disease, or both plants fighting for the same root space.

Companion planting is a popular and effective means of natural insect pest control.

Numerous articles have been written on the subject (see Bibliography).

If companion plants will be compatible to the solarspace as well as the primary plant, they may be a good investment. If your solarspace is home for a cameo rose topiary, however, you may not be willing to have that splendid fragrance intermingled with that of garlic just because the garlic helps repel the pests that attack roses.

The following plant companions may work well in your garden.

| When planting: | Plant with: |
| --- | --- |
| tomatoes | basil: promotes better flavor in crops, repels flies and mosquitoes |
| | bee balm: improves growth and flavor |
| | borage: deters tomato worm |
| | parsley: improves flavor and growth |
| roses | tansy, garlic, and nasturtiums: good general repellents, especially for aphids |
| strawberries | borage: helps repel insects |
| beans and peas | summer savory: improves growth and encourages better flavor in crops, deters bean beetles |
| | rosemary: repels bean beetles |
| | petunias: improve growth |
| | marigolds: repel bean beetles |
| carrots | chives: improve flavor and growth |
| | rosemary: deters carrot flies |
| | sage: repels carrot flies |
| squash | borage: deters leaf worms, improves flavor |
| | nasturtiums: repels squash bugs |
| radishes | chervil: improves flavor |
| | nasturtiums: deter aphids, bugs, and beetles |

| | |
|---|---|
| onions | summer savory: improves flavor |
| | chamomile: improves growth and flavor |
| cabbage | sage: repels cabbage moths |
| | rosemary: repels cabbage moths |
| | peppermint: repels cabbage moths |
| | nasturtiums: deter cabbage moths, aphids, other cabbage pests |
| | chamomile: improves growth and flavor |
| | dill: promotes healthy cabbage, but do not plant near carrots |
| | mint: deters cabbage moths |

6–15. *Snagging white flies with yellow cardboard covered with Tanglefoot (Photo: Darryl J. Strickler)*

Good overall companions that protect against a host of insects as well as promote good general plant health and flavor include: thyme, tansy, nasturtium, tarragon, garlic, marigold, lemon balm, and marjoram. Yarrow enhances oil production of most aromatic herbs. Fennel is disliked by most plants and should be planted in an area of its own. Kohlrabi should be planted away from peas, beans, and tomatoes. (For mail-order information on herbs, see Appendix.)

One method of controlling whitefly pests is with the use of a very sticky substance painted on yellow cardboard. The yellow color attracts whiteflies and the sticky cardboard works like flypaper. It can be staked among the plants or fanned around plants to catch the flies (see fig. 6–15).

Another method for insect control may seem as difficult to live with as the problem itself: using insect predators to control insect pests. Certainly if your solarspace is closely integrated into your home or has limited isolation, this may not be a good choice. Nevertheless, it is effective and bears mentioning.

A very busy pest controller is the amiable ladybug. This little beetle has long been used in outdoor gardens with great success. Though innocent in appearance, this tiny creature devours aphids, scales, and other insect pests, both in adult and larva stages. Two other voracious eaters are the lacewing fly and the praying mantis, each eating hundreds of pests each day. *Trichogramma* wasps control a great variety of pests, although their presence in a solarspace could make them pests themselves.

If your solarspace has ground beds, you might consider moving a fat garden toad inside to make a new home in your beds. It will eat a surprising number of insects.

Another little creature who has a ravenous appetite for pests is the charming chameleon. Frequently sold as pets for children, these gentle little lizards will climb the stems and vines of plants in search of aphids, spiders, and the like. One chameleon could easily patrol a small area of plants. In a larger growing area, two or more may be required to keep the pest population in check. A pair of chameleons may court and bless your space with young ones. (A local pet shop may even be interested in buying the newly hatched babies.)

Some greenhouse gardeners implement an "open door" policy for insect control. It is their belief that by allowing any and all insects to migrate freely, a natural balance will be established. As with the use of insect predators, this may not be a good solution for the well-integrated solarspace.

For other organic controls, check your local garden or farm supply center or a local organic gardening club. There are also publications specific to organic gardening needs (see Bibliography).

## Watering

Watering, a very important part of solar-space gardening, will significantly affect your garden's success. Keep in mind your solarspace environment will make demands on the plants' moisture retaining systems much differently than the indoor or outdoor environment. Plant moisture is robbed from the leaves and soil by the warm air moving from the solarspace on its way into the house. Air returning to the solarspace is much cooler and drier. To protect your plants from becoming too dry, it may be necessary to spray or mist in addition to regular waterings. All sorts of spraying and misting gadgets are for sale (see fig. 6–16)—brass pump sprayers, watering wands that attach to the sink faucet, even automatic watering and misting devices that may be permanently installed (see fig. 6–13). If you have a tank garden sprayer, let it do double duty as a mister in the solarspace. An empty pump-type spray bottle filled with water will mist lots of plants and is very lightweight. The old-fashioned clothes sprinkler does a good job sprinkling leaves and soil and is excellent for watering seed starting trays because it does not splatter the soil or wash away seeds.

Plant moisture should be checked routinely; usually a morning or evening check will suffice unless the solarspace has been very hot or dry. Watering or misting should be done early or late in the day—before 10:00 A.M. or after 4:00 P.M. is a good rule—since the midday sun can easily scorch or steam the life out of a freshly watered plant.

You can make the time you spend spraying really count by mixing in a few drops of liquid soap or other insect repellent. The solution should be up to one-eighth the regular strength because of the buildup that occurs with regular spraying (see formulae and directions in *Pests and Plant Diseases*).

Never—ever—use icy-cold water to spray or water a plant. A cool splash may feel refreshing to you on a sunny day, but your plant will not experience that same relief and may, in fact, go into a thermal shock. Make sure the water is at room temperature, or even a few degrees warmer, before giving it to the plants. Some plants, such as African violets, cannot tolerate

6–16. *Plants in a solarspace need regular watering; watering wand makes the job easier (Pratt solarspace, designed and built by Northern Sun; photo: Darryl J. Strickler)*

having their leaves or blossoms wet, even though they like to be misted. Watering these plants is easier with a long, skinny-spouted watering can or bulb baster from the kitchen, which allow access to the soil without spilling water on the leaves or blossoms.

Regardless of the time, or lack of it, that you have to devote to watering, try to be consistent. Just a few minutes daily for watering and general care will be more productive than several hours spent once a week.

If excessive moisture loss is still a problem, it may be helpful to spray water over the whole bed of soil or rocks, the floor, or even the walls. Of course, watering down the walls or floors is a good idea only if these elements are of a water-proof material, such as masonry or tile, there is sufficient drainage to accommodate the excess water, and the drying conditions of the house and solarspace require extra humidity.

## Plant Grooming

For the price of just a few extra minutes a day, your plants can look terrific. With simple tools grooming can easily be done along with your watering routine. Cotton balls or cotton swabs and rubbing alcohol will quickly clean mealy bugs, scales, aphids, and cobwebs from stems and leaves (see *Pests and Plant Diseases*). Trim brown leaf tips and edges, and straggly plant parts with scissors (trim serrated leaves with pinking shears); loosen the soil surface with an old fork or table knife, being careful not to disturb shallow roots. Velvety leaves can be gently cleaned with a soft toothbrush or complexion brush; smooth leaves should be dusted occasionally with a soft cloth or cotton ball dipped in warm water. Spray leaf glosses can clog leaf pores, but an inexpensive shine is easily produced by wiping the leaves with a small amount of vegetable oil or mayonnaise. Do not leave a thick coating (you are not basting a turkey); use the oil or mayonnaise sparingly, as you would a furniture polish.

## Seed Saving

Seed saving will likely become more important to you each season you grow plants in your solarspace. As you try different varieties of plants, you may notice some types grow and produce better than others. Saving seeds from plants that produce the best crops, color, or foliage will help isolate the strains that grow the best in conditions unique to your solarspace.

Although hybrid seeds will sometimes reproduce, they will not produce plants like the hybrid parent. They are consequently not suitable for seed saving. Check seed catalogs or seed saving groups for stock or standard seeds that will easily reproduce.

Seed saving calls for harvesting, separating the seed from the pulpy part of the plant or pod, drying it away from heat, and storing it in an airtight container. Techniques for seed saving are discussed completely in books listed in the Bibliography.

## Pollination

When produce plants are blooming outdoors, pollination normally is assisted by bees, ants, and other little creatures or by the wind. It may not be such a great idea to open your door to the whole outdoors, but you could assist pollination with a small fan to blow air through the plants. You can also pollinate manually with a cotton swab. Just touch the cotton tip into the center of the flowers of the same type plants, one after another. For example, pollinate all the tomatoes at the same time. Throw away the swabs to prevent spread of disease.

## Straw Bales and Soil Bags

If you want to try something different, try gardening with straw bales or soil bags.

Straw bales take a little time to prepare but are rather easy. The bale should be located on a piece of heavy plastic where drainage has been planned. Soak the bale by pouring water over it. Never let the bale dry out; keep it soaked daily. After about a week or ten days, spread about three handfuls of well-rotted manure on top and water it again. In about two to three weeks, the straw should be decomposing and composted enough in the center of the bale to allow transplanting of a healthy tomato or other plant. (See Bibliography for further reading.)

Soil bags are ready to use without waiting. Place bags of potting soil mixture on top of a couple of boards or bricks to allow for drainage. Poke lots of holes in the bottoms of the bags with a nail or ice pick again, for drainage. On the top of the bag, slice as many Xs with a knife as you have plants to plant. Water occasionally, but remember that the plastic will keep a lot of moisture from evaporating, so overwatering could be a problem. Soil bag gardening will allow you to grow thick banks of plants, yet your garden will remain completely mobile.

## Hydroponics

Hydroponics, a new frontier in gardening that involves growing plants without soil, is a field all its own. Plants are sprouted and then grown in a gravel-type substance lined in troughs (see fig. 6–17). The plants are then flooded several times a day with a nutrient solution. Some say the hydroponic system grows bigger crops faster because the plant does not have to develop such an elaborate root system and can therefore devote more of its energy to fruit or leaf production. Although there are small tabletop of shelf models, the expense of a full-scale hydroponic system should be thoughtfully considered before purchase by the beginning gardener. (See sources listed in Bibliography for further information.)

6-17. *Hydroponic troughs in Weaver solarspace (Photo: Darryl J. Strickler)*

# GROWING WITH YOUR SOLARSPACE

When all is said and done, the most significant rewards you receive from your solarspace are likely to be personal ones. There are benefits that cannot be measured by your thermometers, the amount of food in your refrigerator, your checkbook balance, your income tax refund, or any other quantitative scale, because they have to do with the quality of your life.

The collective experience of hundreds of solarspace owners across America has consistently demonstrated—through formalized surveys and accumulated folklore—that the most important rewards of a solarspace have more to do with the true definition of "living" than with dollars and cents, heating and cooling,

---

### A Greenhouse Fish Farm

Arnie Valdez
People's Alternative Energy Service
San Luis, Colorado

As food and heating costs soar, people are turning to greenhouses to capture solar energy for heat and food production. But even for those who grow fruits and vegetables, a source of low-cost protein continues to be a problem. A partial solution is to use the greenhouse for protein production in the form of fish.

The water-filled tanks used to raise fish can serve as thermal storage mass and as a controlled environment for the fish. In addition, the enriched water in which the fish live can be used to irrigate greenhouse crops. A well-managed aquaculture system will provide a higher yield of protein per pound of feed than land animals can. It will also allow the greenhouse owner to control the climate, nutrients, and water-species factors and build an ecosystem to an extent not possible with conventional agriculture methods.

In many parts of the world, fish farming has been practiced for thousands of years, but here in the United States, it is confined mostly to large-scale commercial operations requiring large amounts of space, capital, and water resources. By the time the product reaches the market, it is a high-cost luxury food. In contrast, the solar greenhouse with aquaculture capabilities max-

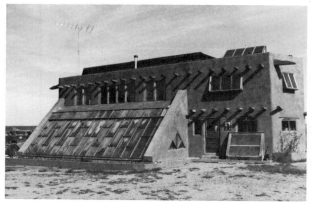

6-18. *Maria and Arnie Valdez's owner-built adobe home (Photo: Darryl J. Strickler)*

imizes the use of available resources at a lower cost.

My wife, Maria, and I had our first introduction to aquaculture when we visited the New Alchemy Institute on Cape Cod. We were inspired by the beauty and harmony of the vegetables and fish growing side by side in New Alchemy's "Ark," a large solar greenhouse. John Todd, one of the originators of the New Alchemy Institute, convinced us of the protein-producing capability of present-day solar greenhouses.

Armed with some basic information, we set out to conduct an aquaculture pilot project in our own greenhouse. The first step was to build fish

*continued ...*

tanks. To make "sun tubes" or "sun silos" for our fish, we rolled flat greenhouse fiberglass glazing into a cylinder 4 feet (1.22 m) long and 18 inches (45.7 cm) in diameter, allowing for a 2-inch (5 cm) overlap. Although our two tanks are only 18 inches by four feet (45.7 cm by 1.22 m), New Alchemy Institute recommends large tanks—up to 5 feet (1.5 m) in diameter. Smaller tanks have larger fluctuations in temperature and pH. The movement of fish are also restricted by the smaller tanks.

First we lightly sanded the overlap, coated it with adhesive, and clamped it with C-clamps holding a 2×4 over the overlapped seam. The clamps were left in place for two days.

After the seam of the cylinder had dried, we cut a circular piece of fiberglass 2 inches (5 cm) larger than the diameter of the cylinder to serve as the tank bottom. The bottom edge of the cylinder and a 1-inch (2.5-cm) band around the bottom were sanded. The seat, or bottom piece, was also sanded where it would contact the bottom of the cylinder. (The sanding assures proper adhesion.) After sanding, a bead of adhesive was applied at the place where the cylinder meets the base. A board with a weight on it was then placed across the top of the cylinder until the adhesive cured. After the adhesive dried (forty-eight hours later) a brush with a handle made from a dowel was used to dab the inside of the bottom seam with adhesive. Another forty-eight-hour period was allowed for the adhesive to harden.

Short pieces of fiberglass tape were used to reinforce the cylinder at the bottom. These were bent in a 90-degree angle to join the bottom of the cylinder to the base, and were held in place by wrapping several layers of fishing line around the tank, over the tape.

The next step was to fill the tanks with water and set them in the sun. We had several leaks the first time and had to drain the tanks, add more adhesive, and recheck. The bottom seal gave us the most trouble. A gallon of water weighs almost 8 pounds (1 liter weighs about 1 kg), so our 18-inch diameter by 4-foot-high (45.7-cm by 1.22-m) tanks contain 50 gallons of water weighing nearly 400 pounds (190 l, about 190 kg). Obviously a sturdy floor is needed to support the concentrated load.

The full tanks should sit in a sunny location for about two weeks to allow plastic residue to be absorbed in the water. Water temperature should be checked at the top, middle, and bottom of the tank. In our tanks temperatures ranged from mid-

6–19. Homemade fish tanks in the Valdez greenhouse, filled with water, algae, and tilapia (Photo: Shirleen Strickler)

seventies to about 82°F (24 to 27.8°C), with temperatures decreasing a few degrees toward the bottom of the tank. In June night temperatures dropped six to eight degrees F.

After the tanks had absorbed the plastic residues, we drained them, rinsed them, refilled them with clean water, and let the temperature stabilize for a few days before introducing the fish.

The *Tilapia aureau*, or blue tilapia, is the type of fish we use. It requires a temperature range of 64 to 90°F (17.8° to 32.2°C), with the best growth at 82° to 86°F (27.8° to 30°C). Tilapia can survive temperatures to about 50°F (10°C), but lower temperatures are fatal. Originally from Africa, tilapia are a very robust fish. They are resistant to disease and can survive low oxygen and high ammonia levels. They grow fast, taste good, and provide an excellent source of protein to supplement greenhouse vegetables. Tilapia are herbivorous, or plant eating. In the solar silo, sunlight enters the water, allowing phytoplankton or algae to develop. This removes ammonia, adds oxygen to the water, and provides a substantial part of the fish diet.

Erwin Young, a commercial fish farmer near

*continued* . . .

Alamosa, Colorado, kindly donated twenty fish ranging in size from 2 to 4 inches. We brought them home in 75°F (23.9°C) water so they would not experience too much shock, placed them in the tank immediately, and turned on a three-way aquarium pump to assure adequate oxygen in the water.

During the first few days, the fish were sluggish and stayed near the bottom of the tank. Every morning they received a sprinkling of floating tropical fish food. The feeding was encouraging. They appeared hungry, coming to the surface for the food. Later we switched to Trout Chow, a commercial fish pellet, which is less expensive than the tropical fish food.

Gradually the water in the tank became green as algae grew. Once the algae had developed to a point where we could not see through the tank, we ceased use of the pumps during the day, because algae in the presence of sunlight provides sufficient oxygen. We still use the pump on cloudy days and at night. After one sunny day, we forgot to turn on the pump at night. Much to our relief, we found all was well the next morning. Apparently enough oxygen was available for the night hours.

Presently the water in the tanks is dark green, with temperatures reaching the high seventies during the day and dropping to the mid-sixties overnight. Every two weeks we drain about a quarter of water from the bottom and use it to water the greenhouse crops. We replace that water with heated water taken from a solar water heater.

We deprive the fish of food pellets one day each week, allowing the Tilapia to feed entirely on algae. In addition to feeding them pellets, we occasionally give them duckweed from a local stream, which they seem to enjoy very much. In the past three months, the fish have doubled in size, with a few approaching 8 inches (20.3 cm). Our stocking density is one fish per 5 gallons (19 l) of water. The New Alchemy Institute has successfully had densities of three fish of various sizes per gallon (3.8 l).

As the fish approach eating size (about 8 inches [20.3 cm]), we will harvest them, keeping the smaller ones through the fall and winter. For the colder weather, we will incorporate a thermostatically controlled immersion heater to ensure the right temperatures for growth when solar heating is not adequate. The heated water also serves as supplemental heating for the greenhouse. With each degree F the temperature rises, a 50-gallon (190-l) tank will gain 400 Btu's of stored energy.

We are really impressed with the results and ease of operating our solar fish farm, but realize that this experience is but a small step toward low-cost protein production using solar energy. Much work and experimentation remains to be done in order to fine-tune the system for our climate and economic conditions. In the coming months, a detailed evaluation of all aspects of this project will be done. We hope to inspire all greenhouse owners to consider a small-scale aquaculture project as a rewarding and localized way of achieving food self-sufficiency.

(The sources listed in the Bibliography provide additional information on aquaculture.)

and food and plants. This is not to suggest that matters pertaining to money, comfort, and horticulture are unimportant, but that most solarspace owners have discovered a wide range of qualitative benefits as well.

Because these benefits are more subjective and intangible, they are not easily described, but they are often conveyed by solarspace owners in direct statements like the following.

*I have to say that I got a real sense of satisfaction out of it. . . . You have to remember . . . I never built anything before. That's one of the things this whole thing [the greenhouse project] has done—convinced me that I could really build something.*

*Actually we didn't plan on growing food—it just sort of happened. Our success—maybe luck—with the first things we planted encouraged us to grow more.*

*. . . being in my sunpace is like getting a warm hug.*

*. . . we spend more time in the greenhouse than anywhere else and people who come to our house always seem to go there first.*

*Most of our meals we eat in the sunspace.*

*I can't seem to keep my kids and cats out of the solarium . . . but then why should I? It's a great place to be.*

*. . . you know, this is a great way to help our children learn about solar . . . firsthand—in their own home.*

*I spend a lot of time in this room [solarspace]— even in summer. . . . It's actually cool in here and cooler in the house since they built my sunspace.*

*We feel more like doing our morning exercises now because we can look out into a space that's filled with plants and light . . . what a way to start the day!*

*. . . since we built the solar addition, we don't go out as much. I guess you could say we're into our hot tub—literally.*

*. . . I haven't got a lot of money, you understand. With this [the greenhouse], I feel like I can make it.*

*A lot of people we know ask us why we built it. When we tell them—and after they've had a chance to sit at our table in the greenhouse, drink a cup of tea, and see all the plants and food growing in the middle of winter—it's not real hard to convince them it was a good idea.*

After you have lived with your solarspace for awhile, you may want to add your comments to those above. When all is said and done, there will always be more to say and do . . . and more to discover. As you grow with your solarspace, enjoy it.

6-20. Seawell solarspace (Designers: Pat Dawe and Josh Comfort; builder: Dan Thane, The Construction, Denver, CO; photo: Darryl J. Strickler)

# Appendix

## SELECTED BIBLIOGRAPHY

Listed below, by category, is a collection of books and articles that are recommended for further reading and reference. (Addresses of publishers are included for titles not readily available from bookstores.)

### Passive Solar Principles and How-to Books

Anderson, Bruce, and Wells, Malcolm. *Passive Solar Energy.* Andover, MA: Brick House Publishing Company, 1981.

Baer, Steven. *Sunspots.* 3rd ed. Albuquerque, NM: Zomeworks Corporation, 1980. (P.O. Box 25806, Albuquerque, NM 87125)

Carter, Joe, ed. *Solarizing Your Present Home.* Emmaus, PA: Rodale Press Inc., 1982.

Mazria, Edward. *The Passive Solar Energy Book.* Emmaus, PA: Rodale Press Inc., 1979.

Reif, Daniel. *Solar Retrofit: Adding Solar To To Your Home.* Andover, MA: Brick House Publishing Company, 1981

Schwolsky, Rick, and Williams, James. *The Builder's Guide to Solar Construction.* New York: McGraw-Hill Inc., 1982.

Strickler, Darryl. *Passive Solar Retrofit: How to Add Natural Heating and Cooling to Your Home.* New York: Van Nostrand Reinhold Company Inc., 1982.

Wright, David. *Natural Solar Architecture: A Passive Primer.* New York: Van Nostrand Reinhold Company Inc., 1978.

Wright, David, and Andrejko, Dennis. *Passive Solar Architecture: Logic and Beauty.* New York: Van Nostrand Reinhold Company Inc., 1982.

### Owner-Builder Design and Construction

McLaughlin, Jack. *The Housebuilding Experience.* New York: Van Nostrand Reinhold Company Inc., 1981.

Owner Builder Center. *Build Your Own House.* Berkeley, CA: Ten Speed Press, 1982.

Walker, Les, and Milstein, Jeff. *Designing Houses: An Illustrated Guide to Building Your Own House.* New York: Overlook Press, 1976.

Wing, Charles. *From the Walls In.* Boston: Little, Brown and Company, 1979.

## Working With Contractors

Heldman, Carl. *Be Your Own House Contractor.* Charlotte, VT: Garden Way Publishing, 1981.

McClintock, Mike. *Getting Your Money's Worth From Contractors.* New York: Harmony/Crown, 1982.

## Construction Techniques

Durbahn, Walter E., and Sundberg, Elmer W. *Fundamentals of Carpentry.* 5th ed. New York: Van Nostrand Reinhold Company Inc., 1982.

Kreh, R. T. *Simplified Masonry Skills.* 2nd ed. New York: Van Nostrand Reinhold Company Inc., 1982.

Love, T. W. *Construction Manual: Concrete and Framework.* Solana Beach, CA: Craftsman Book Company, 1977. (542 Stevens Avenue, Solana Beach, CA 92075)

Love, T. W. *Construction Manual: Rough Carpentry.* Solana Beach, CA: Craftsman Book Company, 1976. (Address above)

*Time-Life Books: Basic Wiring.* Chicago: Time-Life Books.

*Time-Life Books: Masonry.* Chicago: Time-Life Books.

*Time-Life Books: New Living Spaces.* Chicago: Time-Life Books.

*Time-Life Books: Weatherproofing.* Chicago: Time-Life Books.

Van Orman, Halsey. *Illustrated Handbook of Home Construction.* New York: Van Nostrand Reinhold Company Inc., 1982.

Wagner, Willis. *Modern Carpentry: Building Construction Details in Easy-to-Understand Form.* South Holland, IL: Goodheart-Willcox Company, 1976. (123 West Taft Drive, South Holland, IL 60473)

## Solarspace Design, Construction, and Operation

Alward, Ron, and Shapiro, Andy. *Low-Cost Passive Solar Greenhouses: A Design and Construction Guide.* 2nd ed. Butte MT: National Center for Appropriate Technology, 1980. (P.O. Box 3838, Butte, MT 59701)

Craft, Mark, ed. *Winter Greens: Solar Greenhouses for Cold Climates.* Scarborough, Canada: Firefly Books Limited, 1982. (3520 Pharmacy Avenue, Unit 1–C, Scarborough, Ontario, Canada M1W 2T8)

McCullough, James, ed. *The Solar Greenhouse Book.* Emmaus, PA: Rodale Press Inc., 1978.

Yanda, Bill, and Fisher, Rick. *The Food and Heat Producing Solar Greenhouse Book*, 2nd Ed. Santa Fe, NM: John Muir Publications, Inc., 1980.

## Regionally Specific Books

*South*
Novell, Bruce, and Lewis, Kenneth. *The Solar Greenhouse: Design, Construction and Operation for Alabama Homeowners.* Huntsville, AL: Alabama Solar Energy Association, 1981. (ASEC, UA–H, P.O. Box 1247, Huntsville, AL 35899)

*Passive Solar Retrofit Handbook: Solar Applications for Residences.* Atlanta: Southern Solar Energy Center, 1981. (SSEC, 61 Perimeter Park, Atlanta, GA 30341)

Tennessee Valley Authority. *Introduction to Solar Greenhouses: Using the Sun for Home Heating.* Chattanooga, TN: Tennessee Valley Authority, 1981. (TVA, Division of Energy Conservation and Rates, Chattanooga, TN 37401)

*Midwest*
*Proceedings of the Great Lakes Solar Greenhouse Conferences I, II, III, IV and V, 1978–82.* (Proceedings from conferences I, II and V available from: Chippewa Nature Center, 400 South Badour Road, Rt. 9, Midland, MI 48640; proceedings from conferences II, IV available from: Kalamazoo Nature Center, 7000 North Westredge Avenue, Kalamazoo, MI 49007.)

*Northwest*
Magee, Tim et al. *A Solar Greenhouse Guide for the Pacific Northwest.* 2nd ed. Seattle, WA: Ecotope Group, 1979. (2332 East Madison, Seattle, WA 98112)

*West*
Niles, Philip, and Haggard, Kenneth L. *Passive Solar Handbook.* Sacramento, CA: California Energy Commission, 1980. (CEC, Solar Office, 1111 House Avenue, Sacramento, CA 95825)

Snyder, Rachel. *Adding Solar: A Denver Homeowners Guide.* Denver: Denver Solar Energy Association, 1981. (The Solar Bookstore, 1290 Williams Street, Denver, CO 80218)

## Remote Thermal Storage: Rock Beds and Radiant Floors

Lewis, Dan et al. "Rockbeds," *Solar Age,* March, 1982, pp. 44–48.

Lewis, Scott. "Radiant Floors." *Solar Age,* May, 1982, pp. 42–44.

## Air-To-Air Heat Exchangers

Fuller, Winslow. "Installing Household Heat Exchangers." *Solar Age,* September, 1982, pp. 22–23.

Lafavore, Michael. "Clean Air Indoors." *New Shelter,* May/June, 1982, pp. 20–31.

Shurcliff, William. *Air-To-Air Heat Exchangers for Houses.* Andover, MA: Brick House Publishing Company, 1981.

## Batch Water Heaters

Bainbridge, David. *The Integral Passive Solar Water Heater Book.* Davis, CA: The Passive Solar Institute, 1981. (S.U.N., Inc., 450 East Tiffin Street, Bascom, OH 44809)

Gruber, Jim, and Howe, Dick. "A Low-Cost Batch Heater." *Solar Age,* May, 1982, pp. 20–22.

Langua, Frederic. "The Best We Know." *New Shelter,* July/August, 1981, pp. 22–35.

Reif, Daniel. *Passive Solar Water Heaters.* Andover, MA: Brick House Publishing Company, 1982.

## Insulation and Weatherization

Knight, Paul. *Home Retrofitting for Energy Savings.* New York: Van Nostrand Reinhold Company Inc., 1983.

*The Complete Energy Savings Home Improvement Guide.* New York: Arco Publishing, Inc., 1979.

Massachusetts Audubon Society. *The Energy Saver's Handbook for Town and City People.* Emmaus, PA: Rodale Press Inc., 1982.

## Window Insulation

Langdon, William. *Movable Insulation.* Emmaus, PA: Rodale Press Inc., 1980.

Schurcliff, William. *Thermal Shutters and Shades.* Andover, MA: Brick House Publishing Company, 1980.

Wolf, Ray. *Insulating Window Shade.* Emmaus, PA: Rodale Press Inc., 1980.

## Prefabricated Greenhouses and Sunspaces

Lelen, Kenneth. "A Buyer's Guide to Greenhouse Kits." *Solar Age*, May, 1982, pp. 33–40.

## Houseplants and Pest Control

Abrams, George, and Abrams, Kathy. *Organic Gardening Under Glass.* Emmaus, PA: Rodale Press Inc., 1975.

Boodley, James W. *The Commercial Greenhouse Handbook.* New York: Van Nostrand Reinhold Company Inc., 1981.

Crockett, James. *Crockett's Victory Garden.* Boston: Little, Brown and Co., 1977.

Cruso, Thalassa. *Making Vegetables Grow.* New York: Van Nostrand Reinhold Company, Inc., 1975.

DeKorne, James. *The Survival Greenhouse.* El Rito, NM: The Walden Foundation, 1975. (P.O. Box 5, El Rito, NM 87530)

Fisher, George. *Insect Pests.* New York: Golden Press, 1966.

Fisher, Rick, and Yanda, Bill. *The Food and Heat Producing Solar Greenhouse.* Santa Fe, NM: John Muir Publications, 1980.

Halpin, Anne, ed. *Rodale's Encyclopedia of Indoor Gardening.* Emmaus, PA: Rodale Press Inc., 1980.

Hill, Lewis. *Fruits and Berries for the Home Garden.* Charlotte, VT: Garden Way Publishing Company, 1980.

Klein, Miriam, ed. *Biological Management of Passive Solar Greenhouses.* Butte, MT: National Center for Appropriate Technology, 1979. (P.O. Box 3838, Butte, MT 59701)

Klein, Miriam. *Horticultural Management of Solar Greenhouses in the Northeast.* Newport, VT: The Memphremagog Group, 1980. (P.O. Box 456, Newport, VT 05855)

Long Branch Environmental Education Center. *Sun Food: An Introduction to the Production and Use of An Attached Solar Greenhouse.* Leicester, NC: Long Branch Environmental Education Center, 1981. (Route 2, Box 132, Leicester, NC 28748)

Nearing, Helen, and Nearing, Scott. *Building and Using Our Sun Heated Greenhouse.* Charlotte, VT: Garden Way Publishing Co., 1977.

Pierce, John. *Home Solar Gardening.* New York: Van Nostrand Reinhold Company Inc., 1982.

Poincelot, Raymond. *Gardening Indoors with House Plants.* Emmaus, PA: Rodale Press Inc., 1974.

Riotte, Louise. *The Complete Guide to Growing Berries and Grapes.* Charlotte, VT: Garden Way Publishing Company, 1979.

Rodale, J. I. *Encyclopedia of Organic Gardening.* Emmaus, PA: Rodale Press Inc., 1959.

Smith, Shane. *The Bountiful Solar Greenhouse.* Santa Fe, NM: John Muir Publications, 1982.

Westcott, Cynthia. *Plant Disease Handbook.* 3rd ed. New York: Van Nostrand Reinhold Company Inc., 1971.

Wolfe, Delores. *Growing Food in Solar Greenhouses: A Month-by-Month Guide to Raising Vegetables Fruits and Herbs Under Glass.* Garden City, NY: Doubleday and Company, Inc., 1981.

Wylie, Jim. *Attached Solar Greenhouses for Food and Fuel.* Detroit Lakes, MN: Area Vocational Technical Institute, 1981. (Area Vo-Tech, Detroit Lakes, MN 56501)

Zim, Herbert, and Cottam, Clarence. *Insects.* New York: Golden Press, 1956.

## Aquaculture and Hydroponics

Head, William, and Splane, Jon. *Fish Farming In Your Solar Greenhouse.* Eugene, OR: Amity Foundation, 1979. (Box 7066, Eugene, OR 97401)

Huke, Robert and Sherwin, Robert. *A Fish and Vegetable Grower for All Seasons.* Norwich, VT: Norwich Publications, 1977. (Box F, Norwich, VT 05055.)

Jones, Lem. *Home Hydroponics . . . and How To Do It!* New York: Crown Publishers, 1977.

Kenyon, Stewart. *Hydroponics for the Home Gardener.* Rev. ed. New York: Van Nostrand Reinhold Company Inc., 1982.

Logsdon, Gene. *Getting Food from Water.* Emmaus, PA: Rodale Press Inc., 1978.

Todd, Nancy. *The Book of the New Alchemists.* New York: E. P. Dutton Inc., 1977.

# SUPPLIES AND EQUIPMENT

The addresses of manufacturers and distributors listed below should be useful for locating supplies and equipment you may need to complete your solarspace. The best strategy is to call the 800 information operator (800–555–1212) to find out if the manufacturer has a toll-free number in service for your area. Then call (or write) the manufacturer to find out how or where you can order the product. (The 800 numbers listed below usually do not service the state in which the manufacturer or supplier is located.)

## Synthetic Glazing

Exolite (double-skinned acrylic or polycarbonate sheets)
CY/RO Industries
Bound Brook, NJ 08805
(201) 356–2000

Sun-Lite (translucent fiberglass-reinforced polymer glazing)
Solar Components Division
P.O. Box 237
Manchester, NH 03105
(603) 668–8186

Filon (translucent fiberglass-reinforced polyester glazing)
Vistron Corporation, Filon Division
12333 Van Ness Avenue
Hawthorne, CA 90250
(213) 757–5141

Flexigard 7410 and 7415 (clear polymeric composite film)
3M Special Enterprises Department
223–2W 3M Center
St. Paul, MN 55101
(612) 733–0306

Tedlar 400 GT (slightly translucent polyvinyl fluoride film)
Du Pont Customer Service Center
Polymer Products Department
Chestnut Run
Wilmington, DE 19898
(302) 774–8700
(800) 441–9494 (except for West)

Woven Poly (translucent triple-laminated high-density polyethylene sheeting; UV-treated)
Northern Greenhouse Sales
Box 42
Neche, ND 58265
(204) 327–5540
(also supplies Poly-Fastener, a two-part channel locking track useful for mounting polyethylene, PVC film, or shade cloth)

## Glazing Gasket Systems

Maxi Seal and-Edge Seal (for synthetic glazing)
Lane Maxwell Enterprises
4303 Rawhide Road
Pueblo, CO 81008
(303) 542–8906

Sun Haus Glazing Gasket system (for patio glass)
Weather Energy Systems, Inc.
39 Barlows Landing Road
Pocasset, MA 02559
(617) 563–9337

## Operable Roof Windows and Skylights

Velux Roof Windows
Velux-America, Inc.
6180 Atlantic Boulevard
Suite A
Norcross, GA 30071
(404) 448-4551

902 Morse Avenue
Schaumburg, IL 60193
(312) 894-1002

3520 Progress Drive
I-95 Industrial Center
Cornwells Heights, PA 19020
(215) 245-1140

4725 Nautilus Court South
Boulder, CO 80301
(303) 530-1698

Pella Skylight
Rolscreen Company
Pella, IA 50219
(515) 628-1000

Kennedy Sky-Lites (Models V and HPV)
Kenergy Corporation
3647 All American Boulevard
Orlando, FL 32810
(305) 293-3880

Ventarama Skylights
Ventarama Skylight Corporation
75 Channel Drive
Port Washington, NY 11050
(516) 883-5000

Wasco Thermalized Skywindow
Wasco Products, Inc.
P.O. Box 351
Sanford, ME 04073
(207) 324-8060

## Wood-Frame Windows and Sliding Glass Doors

Anderson Corporation
Bayport, MN 55003
(612) 439-5150

Crestline
910 Cleaveland Avenue
Wausau, WI 54401
(715) 845-1161

Marvin Windows
Warroad, MN 56763
(218) 386-1430

Pella Windows
Rolscreen Company
Pella, IA 50219
(515) 628-1000

Sunflake
P.O. Box 28
325 Mill Street
Bayfield, CO 81122
(303) 884-9546

Weather Shield Mfg., Inc.
Medford, WI 54451
(715) 748-2100

## Swing-Open Patio Doors

Swing Set
C-E Morgan, Inc.
P.O. Box 2446
Oshkosh, WI 54901
(414) 235-7170

The Atrium Door
Moulding Products, Inc.
2100 Union Bower Road
Irving, TX 75061
(214) 438-2441
(800) 527-5249 (except Texas)

## Thermal Storage

### Integral Waterwall Containers
*Stud Space Module*
One Design, Inc.
Mountain Falls Route
Winchester, VA 22601
(703) 877-2172

Heatwall
Sun Craft
5001 East 59th Street
Kansas City, MO 64130
(816) 333-2100

Tankwall
Waterwall Engineering
Route 1, Box 6
New Paris, OH 45347
(513) 437-7261

### Phase Change Materials

**RODS**

Clear Heat Energy Storage Tubes
Suncraft Limited
P.O. Box 236
Hinesburg, VT 05461
(802) 482–2163

Thermol 81—The Energy Rod
PSI Energy Systems, Inc.
15331 Fen Park Drive
St. Louis (Fenton), MO 63026
(314) 343–7666

**PODS AND TRAYS**

Kalwall Solar-Pod
Kalwall Corporation
Solar Components Division
P.O. Box 237
Manchester, NH 03105
(603) 668–8186

Concentrated Heat Energy Cells
Solar Research Associates
P.O. Box 3120
Yakima, WA 98903
(509) 248–4817

Thermo-Phase
Holiday Energy Products
600 East Wabash
P.O. Box 465
Waharusa, IN 46573
(219) 862–7224

## Window Insulation

### Roll-down Insulating and Radiant Shades and Curtains

Insulating Curtain Wall, Super Shade, and Thermocell
Thermal Technology Corporation
600 Alter Street
Broomfield, CO 80020
(800) 525–8698 (except Colorado)

Window Comforter
Blue Stem Energy Co-op
1311 Prairie Avenue
Lawrence, KS 66044
(913) 842–7666

Window Quilt Insulating Window Shades
Appropriate Technology Corporation
P.O. Box 975
Brattleboro, VT 05301
(802) 257–4501

Warm Window Insulated Roman Shades
Warm Window
8288 Lake City Way, N.E.
Seattle, WA 98115
(206) 527–5094

Archimedes Energy Shield (Radiant Shade)
Suncraft Limited
P.O. Box 236
Hinesburg, VT 05461
(802) 482–2163

### Insulating Louvers for Skylights

Skylids
Zomeworks Corporation
P.O. Box 25806
Albuquerque, NM 87125
(505) 242–5354

### Homemade Thermal and Radiant Shades—Materials Suppliers

**PULL–DOWN SHADES (ROLLER BLINDS AND RADIANT CURTAINS)**

Foylon 7192 and Dura-Shade 4413
Duracote Corporation
350 North Diamond Street
Ravenna, OH 44266
(216) 296–9600
(800) 321–2252 (except Ohio)

Wind-N-Sun Shield Roller Shades (fabric, rollers, mounting hardware, and track-sealing system)
Wind-N-Sun Shield, Inc.
1793 South Patrick Drive
Indian Harbor Beach, FL 32937
(305) 777–3558

Vapo-Brite and PARSEC Airtight-Brite (metalized film with polyolefin backing; ask for unlabeled material)
PARSEC, Inc.
P.O. Box 38534
Dallas, TX 75238
(214) 324–2741
(800) 527–3454 (except Texas)

Astrolon
   Distributed by:
   Shelter Institute
   38 Center Street
   Bath, ME 04530
   (207) 442-7938

   Manufactured by:
   King-Seeley Thermas Company
   Metallized Products Division
   37 East Street
   Winchester, MA 01890
   (617) 729-8300

### LINERS FOR PSDS INSULATING CURTAINS AND ROLL-DOWN SHADES
Metallized polyethylene and aluminized Mylar
   Shelter Institute
   38 Center Street
   Bath, ME 04530
   (207) 442-7938

Tyvek (spunbonded polyolefin; air-infiltration barrier)
   Du Pont Company
   Spunbonded Products Marketing
   Center Road Building G 123
   Wilmington, DE 19898
   For the nearest dealer call Sweet's Buy-line:
   (800) 447-1980
   (800) 322-4410 (from Illinois)

Airtight-White (Tyvek; ask for unlabeled material)
   PARSEC, Inc.
   P.O. Box 38534
   Dallas, TX 75238
   (214) 324-2741
   (800) 527-3454 (except Texas)

### EDGE SEALS
Magnetic Tape and Steel Tape (first try your local discount department store)
   Shelter Institute
   38 Center Street
   Bath, ME 04530
   (207) 442-7938

   Brookstone Company
   127 Vose Farm Road
   Peterborough, NH 03458
   (603) 924-9511

### TRACK SEALING SYSTEM
Wind Stop
   Wind-N-Sun Shield, Inc.
   1793 South Patrick Drive
   Indian Harbor Beach, FL 32937
   (305) 777-3558

### BUBBLE PACK
Air Cap
   Sealed Air Corporation
   30 West End Road
   Totowa, NJ 07512
   (201) 785-4070

Vikalite (adhesive-backed bubble plastic)
   Viking Energy Systems Co.
   275 Circuit Street
   Hanover, MA 02339
   (617) 871-3180

Foil-Ray (bubble plastic laminated to aluminum foil)
   E. S. I. Company
   8611 West 71st Circle
   Arvada, CO 80004
   (303) 425-6116

### AUTOMATIC SHADE ROLLERS
   Solar Roller Corporation
   709 Spruce Street
   Aspen, CO 81611
   (303) 920-1111

## Components for Solar Water Heaters

### *Tanks for Batch Water Heaters*
   American Appliance
   2341 Michigan Avenue
   Santa Monica, CA 90404
   (203) 829-1755

   Solar Components Corporation
   P.O. Box 237
   Manchester, NH 03105
   (603) 668-8186

*Solarspace on New Mexico residence (Designer-builder: Robert Gibbens; photo: Darryl Strickler)*

### Selective Surface Coatings

Sunsponge
  Berry Solar Products
  2850 Woodbridge Avenue
  Edison, NJ 08837
  (201) 549-3800
  (800) 526-7600

Maxorb Foil and Coatings
  Manufactured by:
  Ergenics
  681 Lawlins Road
  Wyckoff, NJ 07481
  (201) 891-9103

  Distributed by:
  S.U.N., Inc.
  450 East Tiffin Street
  Bascom, OH 44809
  (419) 937-2226
  (800) 537-0985 (except Ohio)

Velvethane-Solar/Optical Black Coating
  Cardinal Industrial Finishes
  1329 Potrero Avenue
  South El Monte, CA 91733
  (213) 444-9274

### Collector Fins

Big Fin
  Zomeworks Corporation
  P.O. Box 25806
  Albuquerque, NM 87125
  (505) 242-5354

### Fittings and Valves

  S.U.N., Inc.
  450 East Tiffin Street
  Bascom, OH 44809
  (419) 937-2226
  (800) 537-0985 (except Ohio)

  Solar Components Corporation
  P.O. Box 237
  Manchester, NH 03105
  (603) 668-8186

## Spa and Hot-Tub Heaters

### Solar

SUN-FRE
  Advanced Solar Technology
  12271 Industry
  Garden Grove, CA 92641
  (714) 895-1375

### Wood-fired

  The Snorkel Stove Company
  P.O. Box 20068
  Seattle, WA 98102
  (206) 523-9637

  Agua Heater Corporation
  P.O. Box 815
  Clark, CO 80428
  (303) 879-3908
  (800) 525-2505 (except Colorado)

## Foilpleat Insulation

  Foilpleat Insulation Inc.
  2020 West 139th Street
  Gardena, CA 90249
  (213) 538-9320

## Exterior Insulation Systems for Masonry or Adobe

  Dryvit Outsulation
  Dryvit Systems, Inc.
  420 Lincoln Avenue
  Warwick, RI 02888
  (401) 463-7150

  4843 Milgen Road
  Columbus, GA 31907
  (404) 563-8021

  5850 South 116th West Avenue
  Tulsa, OK 74107
  (918) 245-0216

Settef System
  Compo Industries, Inc.
  Chemical Specialties Group
  125 Roberts Road
  Waltham, MA 02154
  (617) 899-3000

## Foundation Insulation

  Thermocurve
  Thoro Systems Products
  7800 NW 38th Street
  Miami, FL 33166
  (305) 592-2081

# Air Distribution

### Fans and Automated Control Systems

Wesper 2 fan and Plexus 3 Control System
   Weather Energy Systems, Inc.
   39 Barlows Landing Road
   Pocasset, MA 02559
   (617) 563-9337

Vent-Axia, Inc.
P.O. Box 2204
Woburn, MA 01888
(617) 935-4735

### Fans, Controls, Roof Turbines, and Power Roof Vents

W.W. Grainger, Inc.
5959 West Howard Street
Chicago, IL 60648
(312) 647-8900
(outlets in most major cities)

Mack Ventilator Company, Inc.
565 Lincoln Avenue
Saugus, MA 01906
(617) 233-1730

### Self-operating (Nonelectric) Vent Openers

Series 35 Thermal Operator and Solarvent
   Dalen Products, Inc.
   201 Sherlake Drive
   Knoxville, TN 37922
   (615) 690-0050

Heat Motor
   Heat Motors Distributing
   P.O. Box 411
   Fair Oaks, CA 95628
   (916) 967-0859

Thermafor
   Bramen Company, Inc.
   P.O. Box 70
   Salem, MA 01970
   (616) 745-7765

## Air-to-Air Heat Exchangers

Brener International Corporation
12 Sixth Road
Woburn, MA 01801
(617) 933-2180

Des Champs Laboratories
P.O. Box 440
East Hanover, NJ 07936
(201) 884-1460

Mitsubishi Electric Sales
3030 East Victoria Street
Compton, CA 90221
(213) 537-7132

## Shade Cloth, Solar Screening, and Louvered Screens

Phiferglass SunScreen
   Phifer Wire Products, Inc.
   P.O. Box 1700
   Tuscaloosa, AL 35403
   (800) 633-5955

Phifer Western
14408 East Nelson Avenue
City of Industry, CA 91744
(213) 968-0587

Screen frame for Phifer SunScreen
   Vulcan Metal Products, Inc.
   P.O. Box 6788
   Birmingham, AL 35210
   (205) 956-2000
   (shown in fig. 6-11)

VIMCO Solar Shields (with tension mounting system)
   Virginia Iron and Metal Company
   P.O. Box 8229
   Richmond, VA 23226
   (804) 266-9638
   (800) 446-1503 (except Virginia)

Textilene
   Unitex East
   501 Roosevelt Avenue
   Pawtucket, RI 02863
   (401) 723-6000
   (800) 556-7254 (except Rhode Island)

Unitex West
1641 North Allesandro Street
Los Angeles, CA 90026
(213) 483-9600

Unitex Southwest
609 North Great Southwest Parkway
Arlington, TX 76011
(817) 265-5301
(800) 433-5000 (except Texas)

Louvered screens
Kool Shade Corporation
P.O. Box 210
Solana Beach, CA 92075
(714) 755-5126

Kaiser Shade Screen
Kaiser Aluminum
300 Lakeside Drive
Oakland, CA 94612
(415) 271-3300

Shade cloth and screen is also available from Sears and J.C. Penney.

## Radiant and Vapor Barriers; Air-Infiltration Barrier

Thermosol-Brite, Thermo-Brite, Vapo-Brite, Airtight-White and Thermo-Brite Aluminized Mylar Tape
PARSEC, Inc.
P.O. Box 38534
Dallas, TX 75238
(214) 324-2741
(800) 527-3454 (except Texas)

## DO-IT-YOURSELF PLANS

## Solarspace Plans

Jeff Milstein's Solarspace
Jeff Milstein
P.O. Box 413, Dept. N
Bearsville, NY 12409
(shown in figs. 1-21 and 1-22; $11.00 ppd.; allow 4-6 weeks for delivery)

Sunflake Sun Room
Feeney Associates
Box 406
Durango, CO 81301
($15.00 ppd.)

"Easy to Build" construction manuals for sunspace, batch water heater, and other projects.
T.E.A. Foundation
P.O. Box 29
Harrisville, NH 03450

## Breadbox (Batch) Water Heater Plans

Integral Design
3825 Sebastopol Road
Santa Rosa, CA 95401

Mother's Plans
The Mother Earth News
P.O. Box 70
Hendersonville, NC 28739

Solstice Publications
P.O. Box 2043
Evergreen, CO 80439

Sunspace, Inc.
P.O. Box 172
Ada, OK 74820

Tennessee Valley Authority
Solar Applications Branch
401 Building Annex
Chattanooga, TN 37401
(Do-it-Yourself Solar Water Heater)

Zomeworks Corporation
P.O. Box 25806
Albuquerque, NM 87125

## OWNER-BUILDER SCHOOLS

Cornerstones
54 Cumberland Street
Brunswick, ME 04011
(207) 729-0540

Ecotope
2322 East Madison
Seattle, WA 98112
(206) 322-3995

Going Solar
216 Canyon Acres Drive
Laguna Beach, CA 92651
(714) 494-9341

Heartwood Owner-Builder School
Johnson Road
Washington, MA 01235
(413) 623-6677

Northern Owner Builder
RD #1
Plainfield, VT 05667
(802) 454-7808

Owner Builder Center
1824 Fourth Street
Berkeley, CA 94710
(415) 848-5951

Shelter Institute
38 Center Street
Bath, ME 04530
(207) 442-7938

Sunrise Builders School
Route 121
Grafton, VT 05146
(802) 843-2285

# GROWER'S SUPPLIES

Listed below are addresses of companies that offer such horticultural supplies as seeds, plants, nursery stock, tools, fertilizers, and insect controls. Also listed are distributors of supplies of hydroponic and aquaculture equipment.

## Horticultural Tools and General Supplies

A. M. Leonard, Inc.
6665 Spiker Road
Piqua, OH 45356

Bernard J. Greeson
3548 North Cramer Street
Milwaukee, WI 53211

# Seed Companies and Nurseries

### General Seeds and Stock

W. Atlee Burpee
Box B-2001
Clinton, IA 52732
(seeds, plants, bulbs, nursery stock, dwarf fruit trees, tools, aids, fertilizer, insect predators)

J. A. Demonchaux Company
827 North Kansas Avenue
Topeka, KS 66608
(some European varieties available)

Farmer Seed and Nursery Company
Farivault, MN 55021
(seeds, cold-hardy plants, dwarf fruit trees, berries)

Henry Field Seed and Nursery Company
Shenandoah, IA 51601
(seeds, plants, nursery stock, dwarf fruit trees)

Gurney's Seed and Nursery Company
Yankton, SD 57078
(wide variety of seeds, plants, bulbs, rare and exotic plants, berries, dwarf fruit trees, supplies)

Herbert Brothers Seedmen, Inc.
100 North Main Street
Brewster, NY 10509
(seeds and plants, oriental vegetables, cold-hardy plants)

Johnny's Select Seeds
Albion, ME 04910
(variety of greenhouse-adapted crops)

Nichol's Garden Nursery
1190 North Pacific Highway
Albany, OR 97321
(plants, herbs)

Park's Seed Company
Greenwood, SC 29646
(warm-weather and greenhouse-adapted seeds)

Stern's Nurseries
Geneva, NY 14456
(nursery stock, seeds)

Stoke's Seeds, Inc.
Buffalo, NY 14240
(seeds, cold-hardy plants, greenhouse-adapted crops, oriental vegetables)

Thompson and Morgan
Box 100
Farmingdale, NJ 07727
(seeds, cold-hardy crops, some European varieties)

## Dwarf Fruit Trees and Berries

(in addition to those listed above)
Ideal Fruit Farm and Nursery
Stilwell, OK 74960
(grapes, strawberries, berries, warm-weather-adapted plants)

Lundy's Nursery
Rt. 3, Box 35
Live Oak, FL 32060
(dwarf fruit trees, warm-weather-adapted plants)

Makielsky Berry Farms
7130 Platt Road
Ypsilanti, MI 48197
(strawberries and other berries)

Ozark Nursery
Tahlequah, OK 74464
(strawberries, berries, warm-weather-adapted plants)

Shanti Gardens
Pettigrew, AR 72752
(blueberries—early, midseason, and late varieties)

Stark Brothers Nurseries and Orchard Company
Box A–3424–B
Louisiana, MO 63353
(dwarf fruit trees, roses, ornamentals)

Wolfe Nursery
Highway 377 West
Stephenville, TX 76401
(berries, warm-weather-adapted plants)

## Herbs

(in addition to those listed above)
Calument Herb Company
P.O. Box 248
South Holland, IL 60473
(herbs)

Chientan and Company
1001 South Alvarado Street
Los Angeles, CA 90006
(herbs and oriental vegetables)

Foxhill
Box 70G12
Parma, MI 49269
(culinary, medicinal, and dye herbs)

Stillcopper Herb Farm
Box 186–04
Brookneal, VA 24528
(culinary, medicinal, dye, fragrant, and everlasting herbs)

Werner's Herbs
Route 1, Box 71
Letohatchee, AL 36047
(potted herb plants)

## Oriental Plants and Vegetables

(in addition to those listed above)
Kitasawa
356 West Taylor
San Jose, CA 95110

R. H. Shumway
628 Cedar
Rockford, IL 61101

Tsang and Ma International
P.O. Box 294
Belmont, CA 94002

## Cacti and Succulents

K & L Cactus Nursery
12712 Stockton Boulevard
Galt, CA 95632

Ed Storms, Inc.
4223 Pershing
Fort Worth, TX 76107

*Story-and-a-half solarspace on Francisco residence
(Designer: Rainshadow Design; photo: Darryl Strickler)*

## Rare, Exotic, Traditional, and Heirloom Seeds and Plants

Graham Center Seed Directory
Route 3, Box 95
Wadesboro, NC 28170
(catalog of small, traditional seed companies)

International Growers Exchange
Box 397
Farmington, MI 48024
(rare and exotic plants)

Rod McLellan Company
1455 El Camino Real
South San Francisco, CA 94080
(orchids)

Siskiyou Rare Plant Nursery
522 Franquette Street
Medford, OR 97501
(rare plants)

True Seed Exchange
Route 1
Princeton, MO 64673
(heirloom and scarce traditional seeds)

Yamada Nursery
1395 Ainaola Drive
Hilo, HI 96720
(anthurium and other exotic plants)

## Plants and Seeds Available Outside the United States

Hertel Gagnon
R.R. 3
Compton, Quebec
Canada
(dwarf fruit trees)

Mr. Gardener, Inc.
Niagara Stone Road
Virgil, Ontario
Canada
(dwarf fruit trees, plants)

Richters
Goodwood, Ontario, L0C1A0
Canada
(seeds, herbs)

Sutton Seeds Ltd.
London Road, Earley
Reading, Berkshire RG6 1AB
England
(seeds and plants)

## Fertilizers

Atlas Fish Emulsion Fertilizer
Menlo Park, CA 94025
(fish emulsion)

Charles Bateman, Ltd.
135 Highway 7 East
Thornhill, Ontario
Canada
(fish emulsion high in phosphorus and potassium)

Green Earth Organics
9422 144 Street East
Puyallup, WA 98371
(organic fertilizers, including seaweed and kelpmeal)

Organic Farm and Garden Supply
115 Warner Drive
Columbia, SC 29204
(organic fertilizers and supplies)

United States Organic Ferto Corporation
P.O. Box 111
Spanish Fork, UT 84660
(fertilizers)

## Insect Controls

Better Yield Insects
13310 Riverside Drive East
Tecumseh, Ontario N8N 1B2
Canada
(insect predators; orders for United States must be accompanied with a U.S.D.A. permit obtainable by writing: U.S.D.A. Permit Unit, Room 635, Federal Bldg., Hyattsville, MD 20782)

W. Atlee Burpee
Box B–2001
Clinton, IA 52732
(ladybugs, praying mantises)

Down-to-Earth Environmental Products
141 East 26th Street
Holland, MI 49423
(diatomaceous earth)

Fountain Sierra Bug Company
Ladybug Lane, P.O. Box 114
Rough And Ready, CA 95975
(ladybugs)

Nature's Control
2518 Stewart
Medford, OR 97501
(spider mite predators)

Rincon-Vitova Insectories, Inc.
P.O. Box 95
Oak View, CA 93022
(insect predators)

White Fly Control Company
Box 986
Milpitas, CA 95035
(whitefly predators)

## Greenhouse Insurance

Florists' Mutual
500 St. Louis Street
Edwardsville, IL 62025

## Hydroponic Supplies

Aqua-Ponics
17221 East 17th Street
Santa Ana, CA 92701

Artex Hydroponics
Mena, AR 71953

Hydro Farms, Inc.
2405 T.O. Harris Road
Box 144–H
Mansfield, TX 76063

Hydroculture Inc.
P.O. Box 1655
Glendale, AZ 85301

Robert B. Peters Company, Inc.
2833 Pennsylvania Street
Allentown, PA 18105

Specialty Gardens, Limited
90 Earlton Road
Agincourt, Ontario M1T 2R6
Canada

## Fish Culture Supplies

Keeton Fisheries
Box 359
Laporte, CO 80535

Fish Breeders of California Inc.
Box 1004
Niland, CA 92257
(619) 348–0547

Southern Fish Cultivists, Inc.
P.O. Box 251
Leesburg, FL 32748
(904) 787–1360

## METRIC CONVERSIONS

**1 inch = 2.5 centimeters**
**1 foot = 0.3 meters**
**1 square foot = 0.09 centares = 0.09 square meters**
**1 cubic foot = 0.03 steres = 0.03 cubic meters**
**1 gallon = 3.8 liters**
**1 pound = 0.5 kilograms**

*Low-cost solarspace in Pennsylvania constructed from recycled materials to harmonize with older house (Photo: David Strickler, Strix Pix)*

# Acknowledgments

It is almost impossible to watch an Academy Awards ceremony without hearing one famous person after another refer to "all the wonderful people who made it all possible." And although no one is likely to offer to buy the movie rights to this book, nor am I what you would call a famous person, it is no less appropriate to sincerely thank the numerous individuals who helped to make this book a reality.

In traveling across this land by diesel auto and telephone to gather the material for this book, I was indeed fortunate to talk with a great many outstanding people who cared enough to share what are certainly among their most valuable possessions—their time and talents. These people include:

■ the professional designers, builders, and other individuals whose works have influenced this author and this book:

| | |
|---|---|
| Blair Abee | Bruce Bolme |
| Steve Andrews | Bonnie Bruce |
| Dale Aspevig | John Burton |
| David Bainbridge | M. R. Carey |
| David Baylon | Dale Clark |
| Nancy Benner | Jeremy Coleman |
| Rob Bishop | Terry Egnor |
| Kent Briddell | Rollin Francisco |
| Charles Bliege | Paul Gallimore |

| | |
|---|---|
| Fred Gant | Dale Osborn |
| Antonio Garcia | Larry Price |
| Robert Gibbens | Leon Ragsdale |
| Sally Goble | Mark Randall |
| Arden Handshy | Jeff Reiss |
| Mike Hedrick | Dave Roberts |
| Alexa Johnson | Don Schramm |
| Jim Johnson | Paul Shippee |
| Akira Kawanabe | Bob Skilton |
| Steve Kenin | Eric Sorenson |
| Rob Lerner | Robert Spiker |
| Jerry Liveoak | Robin Taylor |
| Charles Loesch | Angela Teinan |
| Stephen Merdler | Arnold Valdez |
| Tim Magee | Maria Valdez |
| Joe Mestas | Rip Van Winkle |
| Gene Metz | Mike Verwey |
| M. S. Milliner | Valerie Walsh |
| Jeff Milstein | Jim Wiley |
| Lynn Nelson | Harriet Wise |
| David Norton | David Wright |
| Bruce Novell | Peggy Wrenn |
| Paul Olson | Bill Yanda |
| Kenneth Olson | Andy Zaugg |

■ the hundreds of good people who opened their doors and their lives to a weary traveler in search of the best solarspaces in America. I especially want to thank the following individuals, although I am no less grateful to the people I visited whose names do not appear below because of space limitations:

| | |
|---|---|
| Eric Asendorf | Riccardo Pacini |
| James Baker | Sylvia Pratt |
| Joyce Baker | Harry Reiley |
| Leon Brauner | Ann Rodgers |
| Roberta Brauner | John Rodgers |
| James Cason | Eileen Ross |
| Joada Crawford | Robert Ross |
| Sheila Daar | Elizabeth Saccone |
| Charles Dunnell | Glenn Saccone |
| Betty Flynn | Barbara Samuel |
| Robert Flynn | Richard Scarlett |
| Carl Gilmore | Rita Scarlett |
| Elizabeth Gilmore | Buie Seawell |
| Tom Hard | Marjorie Seawell |
| Laurie Harrison | Ethel Smith |
| Doug Kelly | Don Stockton |
| Kathy Kelly | Judy Stockton |
| Tom Lamm | Clifford Surrett |
| Diane Meier | Denzil Surrett |
| Gary Meier | Teresa Vigil |
| Glenda Mitchell | Victor Vigil |
| Martha Mortensen | Jackie Weiss |
| Nancy Nelson | Vern Weiss |
| Regina Pacini | Roslyn Willis |

■ the talented people who contributed their excellent photographs to help tell the visual story in this book:

| | |
|---|---|
| Leon Brauner | Bruce T. Novell |
| Robert Flynn | John Rodgers |
| Paul Gallimore | Don Schramm |
| Jerry Liveoak | David S. Strickler |
| James Marlinski | Shirleen C. Strickler |
| Jeff Milstein | Maria Valdez |
| David Norton | Valerie Walsh |
| | Harriet Wise |

A special debt of gratitude is owed to:

■ Barbara Ravage, Senior Editor, Van Nostrand Reinhold Company, who provided the faith, trust, and good advice I needed to translate the experience gained from years of work, 37,000 miles of travel, and visits to hundreds of solarspace owners and designers into the words and illustrations that became this book. Her expert but gentle guidance, moral support, and fine-tuning of the manuscript are sincerely appreciated and gratefully acknowledged.

■ Linda Rathjen, Van Nostrand Reinhold Company, to whom I am doubly indebted since I failed to adequately acknowledge her outstanding contributions to my last book. I especially appreciate the "English teacher" in her—the part of her professional self that will not allow the careless high school freshman in me to get by with only a "gentleman's C." I also greatly admire her ability to replace five words with one in a single sweep of a red pencil, making it all sound a whole lot better.

■ Winona Henderson, a first-rate person in every sense, who patiently and carefully typed, retyped (through no fault of her own), and proofread the entire manuscript for this book. Her excellent work, uncomplaining attitude, and continued friendship through it all are sincerely appreciated. She accomplished in a few months what would have taken me years if left to my "Hunt-Peck-Swear-and-Tear" approach to typing.

■ Maria and Arnold Valdez, People's Alternative Energy Service, San Luis, Colorado, who have, through their work and through their example, provided valuable inspiration to me and thousands of others, helping to ensure the promise of a brighter energy future for us all. I especially want to thank Arnold for the excellent article he wrote on aquaculture that appears in chapter 6.

■ Cheryl and Darrel Watkins, who graciously allowed me and a crew of Dutchmen to make a royal mess out of their yard, tear their house apart, build them a solarspace, and record it all in the photos that appear in chapter 5 —and then paid for it all.

Finally, and most importantly, I want to thank the person who contributed the most to the completion of this book, my friend, fellow traveler, counselor, partner, lover, and wife, Shirleen Cummins Strickler. Her contributions to this book, and to me, are numerous and significant. Not only did we travel across America together to "get the story," but she also shot many of the more inspired photographs that appear in the book, and carefully researched and singlehandedly wrote the section of chapter 6 that deals with growing food and houseplants. Thanks Sheen: you and the rest of the world know I couldn't have done it without you.

# Index